POWER ENGINEERING, CONTROL AND INFORMATION TECHNOLOGIES IN
GEOTECHNICAL SYSTEMS

2015 ANNUAL PUBLICATION

Power Engineering, Control and Information Technologies in Geotechnical Systems

Editors

Genadiy Pivnyak
Rector of the State Higher Educational Institution "National Mining University", Ukraine

Oleksandr Beshta
Vice Rector (Science) of the State Higher Educational Institution "National Mining University", Ukraine

Mykhaylo Alekseyev
Dean of Information Technology Department of the State Higher Educational Institution "National Mining University", Ukraine

CRC Press
Taylor & Francis Group
Boca Raton London New York

CRC Press is an imprint of the
Taylor & Francis Group, an **informa** business
A BALKEMA BOOK

First published 2015 by CRC Press/Balkema

Published 2019 by CRC Press
Taylor & Francis Group
6000 Broken Sound Parkway NW, Suite 300
Boca Raton, FL 33487-2742

© 2015 by Taylor & Francis Group, LLC
CRC Press is an imprint of the Taylor & Francis Group, an informa business

First issued in paperback 2019

ISBN 13: 978-0-367-45215-5 (pbk)
ISBN 13: 978-1-138-02804-3 (hbk)

Visit the Taylor & Francis Web site at
http://www.taylorandfrancis.com

and the CRC Press Web site at
http://www.crcpress.com

Typeset by Alyona Khar', State Higher Educational Institution "National Mining University", Dnipropetrovs'k, Ukraine

Power Engineering, Control and Information Technologies in Geotechnical Systems – Pivnyak, Beshta
& Alekseyev (eds)
© 2015 Taylor & Francis Group, London, ISBN 978-1-138-02804-3

Table of contents

Power Engineering, Control and Information Technologies in Geotechnical Systems – Pivnyak, Beshta
& Alekseyev (eds)
© 2015 Taylor & Francis Group, London, ISBN 978-1-138-02804-3

Preface

The book is dedicated to solving important problems of efficiency of geotechnical systems. Written by leading scholars and experts in the field, the book presents new topics in the given area. It covers up-to-date technical information that will help to improve operation of geotechnical systems and their parameters. Great attention is paid to the application of mathematical models and algorithms. The book presents a variety of examples to illustrate how the methods can be used to model physical processes and to find solutions to challenging problems The work is remarkable for wide range of considered problems. Practical questions of geotechnical systems are considered in detail such as mining equipment operation, power supply systems of coal mines, electric vehicles, control algorithms for reduction of specific energy consumption in geotechnical systems and others. Industry specialists and researchers are able to use presented results of research for reducing energy consumption and enhancing parameters of geotechnical systems.

Genadiy Pivnyak
Oleksandr Beshta
Mykhaylo Alekseyev

Power Engineering, Control and Information Technologies in Geotechnical Systems – Pivnyak, Beshta
& Alekseyev (eds)
© 2015 Taylor & Francis Group, London, ISBN 978-1-138-02804-3

Traction and energy characteristics of no-contact electric mining locomotives with AC current thyristor converters

G. Pivnyak, M. Rogoza, Yu. Papaika & A. Lysenko
State Higher Educational Institution "National Mining University", Dnipropetrovs'k, Ukraine

ABSTRACT: Results of research concerning traction and energy characteristics of power circuits of no-contact mining electric transport in term of its use in control circuits of electric locomotive traction drive of AC current thyristor converters with short-circulating effect are represented. Recommendations to select efficient alternative of converter in control circuits are given.

1 INTRODUCTION

Improvement of technological processes for minerals mining is closely connected with technical and economic opportunities of locomotive transport being one of basic transport types in horizontal workings of mines.

Development of rail haulage electric equipment system with inductive power transfer (high frequency AC current transport) applying a principle of high frequency electromagnetic power transfer is one of the most upcoming trends for underground transport progress. That makes it possible to design no-contact mining electric locomotives in terms of present-day developments in the field of electrical engineering, power electronics, and mine-type electrical equipment; the locomotives differ fundamentally from available ones.

Industrial prototypes of such innovative electrical equipment tested in coal mines of Donbass demonstrate significant advantages of no-contact electric locomotives to compare with known types of mining locomotives as well as capabilities and expediency for the development.

2 MAIN PART

One of the key problems of mining transport system with inductive energy transfer is *to develop control system for an electric locomotive traction drive* in terms of semiconductor converters of electromagnetic energy parameters providing efficient traction and energy characteristics for transport electric equipment complex while controlling traction force and velocity relying on results concerning research of electromagnetic processes and its power circuit characteristics.

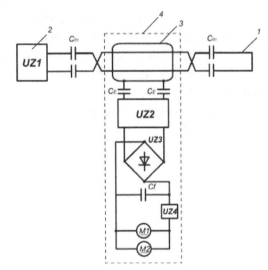

Figure 1. Basic circuit of transport system with inductive power transfer.

A system of no-contact rail transport (Fig.1) implements a principle of high frequency power electromagnetic transfer to mobile objects. Transportation system involves transmitter in the form of electric-traction circuit 1 energized by stationary source 2 – traction thyristor frequency converter. Electric power receiver 3 is located within carriage rolling stock 4 being separated from electric-traction circuit with air isolation. While using high frequency current (e.g. 5 kiloHertz) and obtaining adequate output parameters of traction converter, the system achieves required character of energy conversion to supply drive motors traction M1 and M2 providing its safe environment in coal mines.

Electric locomotive develops its maximum power when receiving circuit is in acceptor resonance of voltage. High Q (-factor) of circuit, current high frequency, and availability of balancing capacitors **Ce** determine traction drive operation specifity as well as peculiarities of adequate controlling mean selection.

To control traction drive, last developments of no-contact electric locomotives use DC voltage pulse width converter UZ4 in a traction drive circuit (Fig.1). However, such a circuit to control traction drive mainly developed for mine battery locomotives has a number of substantial defects while using for no-contact electric locomotives: availability of massive and expensive input filter; adverse effect of voltage pulse control within traction drive terminals on their operation; increase in power loss due to increased number of electric energy parameter transformation phases etc.

Application of no-contact electric locomotive traction drive control circuit in terms of AC voltage semiconductor converters (AC VSC) enables eradicating defects typical for a circuit based upon DC voltage pulse converter. Circuit implementation of such converters should be based on an idea of shunting by means of semiconducting (e.g thyristor) full-wave alternating current switch along with phase control of extra reactive components (e.g. extra reactor) or a group of such elements in electric locomotive AC circuit – converter UZ2 (Fig. 1).

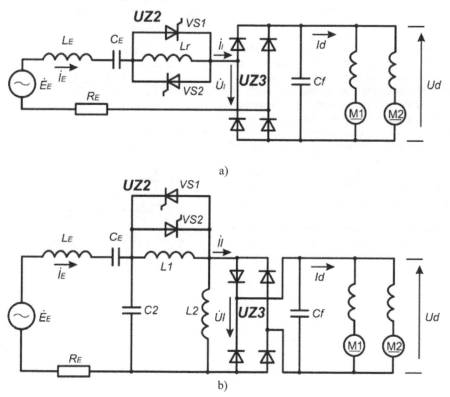

Figure 2. Equivalent circuit of no-contact electric locomotive power circuit with AC semiconductor converters components connection of short-circulating effect.

Circuit implementation of converter which reactive components have flat-topped connection are of the greatest interest (Fig. 2, b). Moreover parameters of the components are selected in such a way that if L1 reactor is completely shunted then L2 – C2 components should form parallel resonance circuit being notch network (filter circuit)) in a power receiver circuit (Pivnyak, Remizov & Saratikyanz 1990). In this context nominal voltage energizes traction drives to be in accordance with traction drive operation mode in terms of standard characteristics. If L1 reactor is introduced (under certain control angle α of UZ2 converter thyristors VS1 and VS2) a new parallel resonance circuit formed by a circuit of traction drives appears (under starting condition it has small resistance practically shunting L2 reactor by C2 condenser and L1 reactor). Heavy reactive current flows in the parallel resonance circuit being simultaneously starting current of traction drives while current of receiving and power circuit IE will be small as parallel resonance circuit L2 – C2 having large active resistance is connected up. Thus, transformation effect of low-resistance traction drive in a starting condition into high resistance of receiving and power circuit takes place. Motor acceleration results in its resistance increase that is shunting action on L2 reactor weakens. Value of equivalent induction L1 decreases; resonance circuit current as well as traction drive current decreases. Such a converter allows reducing sufficiently energy consumption owing to decreased resistance in a traction network; the matter is that in the starting condition current of receiving and power circuit will be less to compare with load current.

While considering such control circuits, characteristics of transport system should be represented depending upon load current and control parameters. In this case such a control parameter is represented by connection angle of VS1 and VS2 thyristors of converter **α**. Load current I*l and* angle **α** are the factors effecting the whole system of no-contact electric transport.

Voltage regulation characteristic of receiving and power circuit of electric locomotive U*l* = f (I*l*, α) is one of the most important characteristics of transport electric equipment exercising a significant influence on traction properties of electric locomotive and energy data.

Electromechanics characteristics of electric locomotive traction drive are plotted on voltage regulation characteristics of receiving and power circuit.

As electric potential to traction drives are applied through full-wave rectifier UZ3 with output filtering capacitor CF, then both current and voltage of DC circuit Ud and Id may be expressed through effective values of receiving and power circuit voltage and current as (Pivnyak, Remizov & Saratikyanz 1990) Ud = 1.11 U*l*, Id = 0.9 I*l*; that is U*l* = f (I*l*, α) dependences provide similar qualitative situation of voltage control within

terminals of traction drives as Ud = f (Id, α) characteristic.

Analytic expressions describing voltage regulation characteristics of receiving and power circuit are basis to analyze characteristics of no-contact electric transport electric equipment; first of all that concerns its energy characteristics.

In terms of topological method of directed graphs, analytic description of electric locomotive receiving and power circuit was obtained for converter with flat-topped component connection (Fig.2,b):

$$\dot{U}_l = \dot{E}_E \left(B_1 - jB_2 \right) - \dot{I}_l \left(B_3 + j\, B_4 \right) \tag{1}$$

where $\dot{E}_E$ is EMF of power receiver;

B_1, B_2, B_3, and B_4 are coefficients involving values of elements of power circuit and α angle of thyristor turn-on:

$$B_1 = \frac{\frac{L_2}{C_2}\left(\frac{2\alpha}{\pi}\frac{L_1}{C_2} + \frac{L_2}{C_2} \right)}{\left(\frac{2\alpha}{\pi}\frac{L_1}{C_2} + \frac{L_2}{C_2} \right)^2 + \frac{4\alpha^2}{\pi^2}\omega^2 L_1^2 R_E^2};$$

$$B_2 = \frac{\frac{2\alpha}{\pi}\omega L_1 R_E \frac{L_2}{C_2}}{\left(\frac{2\alpha}{\pi}\frac{L_1}{C_2} + \frac{L_2}{C_2} \right)^2 + \frac{4\alpha^2}{\pi^2}\omega^2 L_1^2 R_E^2};$$

$$B_3 = \frac{R_E \frac{L_2}{C_2}\left(\frac{2\alpha}{\pi}\frac{L_1}{C_2} + \frac{L_2}{C_2} - \frac{2\alpha}{\pi}\omega^2 L_1 L_2 \right)}{\left(\frac{2\alpha}{\pi}\frac{L_1}{C_2} + \frac{L_2}{C_2} \right)^2 + \frac{4\alpha^2}{\pi^2}\omega^2 L_1^2 R_E^2};$$

$$B_4 = \frac{\frac{2\alpha}{\pi}\frac{L_1}{C_2}\frac{L_2}{C_2}\omega\left(\frac{2\alpha}{\pi}L_1 + L_2 \right) + \frac{2\alpha}{\pi}\omega L_1 R_E^2\left(\frac{2\alpha}{\pi}\omega^2 L_1 L_2 - \frac{L_2}{C_2} \right)}{\left(\frac{2\alpha}{\pi}\frac{L_1}{C_2} + \frac{L_2}{C_2} \right)^2 + \frac{4\alpha^2}{\pi^2}\omega^2 L_1^2 R_E^2}.$$

Fig. 3 demonstrates voltage regulation characteristics of electric locomotive receiving and power circuit for the circuit. Characteristics are represented in relative units as $U_{l*} = f(I_{l*})$ if α = const; $U_{l*} = U_l / E_E$, $I_{l*} = I_l / I_{l.h}$ where $I_{l.h} = 1,11\, I_{d.h}$ is hourly mode current of no-contact electric locomotive traction drives.

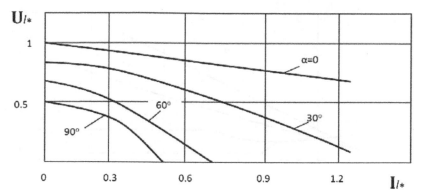

Figure 3. Voltage regulation characteristics of electric locomotive receiving and power circuit for a circuit with flat-topped element connection of AC VSC of shunting effect.

Analysis of voltage regulation characteristics helps to conclude that application control circuit with fat-topped element connection in terms of control demonstrates "autotransformer effect" of voltage excursion within traction drive terminals. That helps to extend greatly a range of voltage control, for example, to compare with a circuit based on AC VSC of shunting action with extra reactor.

Analysis of stiffness factor of voltage regulation characteristic of receiving and power circuit βVR = | dIl / dUl | shows that application of a circuit with flat-topped component connection makes it possible to increase sufficiently (5 to 15 times depending upon terms) their hardness to compare with a circuit involving extra reactor which cuts possibility of electric locomotive slippage.

When control circuit with flat-topped component connection is applied, actual current of receiving and power circuit IE is not equal to actual value of load current Il while controlling. In this case actual current of receiving and power circuit is defined as

$$I_E = \frac{E_E}{\sqrt{R_{el}^2 + X_{el}^2}} \qquad (2)$$

where R_{el}, X_{el} are resistance and reactance of electric locomotive AC circuit respectively:

$$R_{el} = R_E + \frac{R_l}{\left(1 - \frac{2\alpha}{\pi}\frac{L_1}{L_2}\right)^2 + \frac{4\alpha^2}{\pi^2}\omega^2 C_2^2 R_L^2 \frac{L_1^2}{L_2^2}};$$

$$X_{el} = \frac{\frac{2\alpha}{\pi}\omega L_1\left[1 - \omega^2 C_2^2 R_l^2 - \frac{2\alpha}{\pi}\frac{L_1}{L_2}\left(1 + \omega^2 C_2^2 R_l^2\right)\right]}{\left(1 - \frac{2\alpha}{\pi}\frac{L_1}{L_2}\right)^2 + \frac{4\alpha^2}{\pi^2}\omega^2 C_2^2 R_l^2 \frac{L_1^2}{L_2^2}}.$$

Load resistance is defined from equation of voltage regulation characteristic of receiving and power circuit as

$$R_l = \left[\left(\frac{E_E}{I_l}B_1\right)^2 + \left(\frac{E_E}{I_l}B_2\right)^2 - B_4^2\right]^{\frac{1}{2}} - B_3 \qquad (3)$$

Analysis of obtained dependences IE = f (Il , α) (Fig.4) helps to identify that if regulated, control circuit with flat-topped component connection makes it possible to obtain effect of electric locomotive receiving and power circuit current reduction to compare with load current especially if Il > 0.5 …0.6. That provides a means of substantial reduction in electric power consumption under the conditions of traction drive starting and controlling.

4

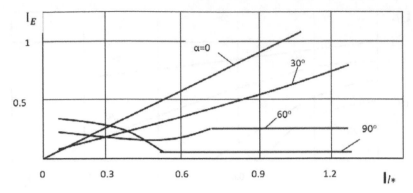

Figure 4. Predicted dependences of receiving and power circuit current on load current while controlling for a circuit with flat-topped component connection of AC VSC shunting action.

Resistance reflected on electric-traction circuit is one of the most important energy characteristics exercising a significant influence on performance indicators of no-contact electric transport. It depends on the fact that no-contact electric locomotives have inductive coupling with electric-traction circuit; moreover, both electric locomotives and electric-traction circuit (contour and balancing capacitors) are load for supply equipment of no-contact electric transport – traction thyristor frequency converter UZ1 (Fig.1).

It is rather important for the latter to know a range of resistance and reactance of electric-traction circuit when one or several electric locomotives operate as effective current of electric-traction circuit can be determined (Pivnyak, Remizov & Saratikyanz 1990) as:

$$I_{tc} = \left. E_{fc} \middle/ \sqrt{(R_{tc} + R_r)^2 + (X_{tc} + X_r)^2} \right. , \qquad (4)$$

where R_{tc}, X_{tc} are resistance and reactance of traction network;

R_r, X_r are resistance and reactance reflected on traction network of no-contact electric locomotives resistance;

E_{fc} is EMF of the traction thyristor frequency converter traction thyristor frequency converter UZ1 (Fig.1).

In addition, variation range of electric-traction circuit resistance and reflected resistance should be known to make adequate choice of circuit and parameters of traction frequency converter which should provide a mode of regulated current within a range of load variation (Pivnyak, Remizov & Saratikyanz 1990) as EMF of electric locomotive consumer (Pivnyak, Remizov & Saratikyanz 1990) is:

$$EE= \omega \, M \, I_{tc}, \qquad (5)$$

where M is mutual inductance of receiving and power circuit and traction network.

Under the assumption of constancy, values M, reflected on electric-traction circuit, resistance and reactance are determined (Pivnyak, Remizov & Saratikyanz 1990) as:

$$R_r = \left(\tfrac{\omega M}{Z_{el}}\right)^2 R_{el} ,$$

$$X_r = -\left(\tfrac{\omega M}{Z_{el}}\right)^2 X_{el} \qquad (6)$$

Where R_{el}, X_{el} are resistance and reactance of receiving and power circuit (no-contact electric locomotive AC circuit); and Z_{el} is a module of no-contact electric locomotive AC circuit complex resistance.

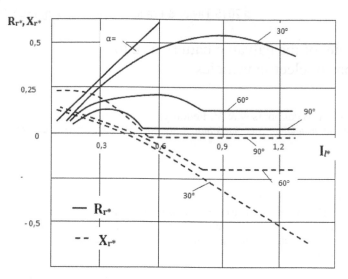

Figure 5. Predicted dependences of load current resistance and reactance reflected on traction network in the process of regulation for a circuit with flat-topped component connection of SC shunting action.

Analysis of resistance reflected of traction network in the process of traction network regulating ($Rr = f(Il, \alpha)$, $Xr = f(Il, \alpha)$ dependences) (Fig.5) demonstrates that application of control circuit with flat-topped component connection makes it possible to decrease resistance reflected on traction network (1.6 to 4 times depending upon terms) to compare with a circuit with extra reactor.

That will help to reduce capacity of thyristor frequency converter, increase its operation reliability, and cut operating costs.

3 CONCLUSIONS

1.   Analysis of voltage regulation characteristics of receiving and power circuit of no-contact electric locomotive with different control circuits of traction drive in terms of AC thyristor converters shows that it is expedient to design such circuits on the basis of AC voltage converter of shunting action with flat-topped component connection.

2.   If control circuit with flat-topped component connection then "autotransformer effect" of voltage excursion within traction drive terminals takes place. That enables considerable extend to voltage control range as well as 5 to 15 times increase in voltage regulation characteristics hardness to compare with a circuit involving extra reactor; it decreases risk of electric locomotive slipping.

3.   Control circuit with flat-topped component connection renders it possible to achieve effect of current reduction for receiving and power circuit of electric locomotive to compare with load current; it helps to cut down energy consumption while regulating.

4.   Level of resistance reflected on traction network while starting and regulating is quite lower (1.6 to 4 times) to compare with a circuit with extra reactor; that will facilitate to cut capacity of traction frequency converter, enhance operational reliability cutting down maintenance expenditures.

5.   Parameter values of control circuit with flat-topped component connection exercise a significant influence on voltage-control range within traction drive terminals, hardness of receiving and power circuit voltage regulation, receiving and power circuit current magnitude, and level of resistance reflected on traction network. Efficient parameters of circuit components are determined on the basis of obtained characteristics analysis. Recommendations are given as for selection of the value for current-limiting setting of control circuit starting mode on the assumption of efficient characteristics providing for no-contact transport electric equipment complex.

REFERENCES

Pivnyak G., Remizov I., Saratikyanz S. et. al.; edited by G. Pivnyak. 1990. *Transport with induction transmission of energy for coal mines* (in Russian). Moscow: Nedra: 245.

Power Engineering, Control and Information Technologies in Geotechnical Systems – Pivnyak, Beshta & Alekseyev (eds)
© 2015 Taylor & Francis Group, London, ISBN 978-1-138-02804-3

Universal model of the galvanic battery as a tool for calculations of electric vehicles

O. Beshta, A. Albu, A. Balakhontsev & V. Fedoreyko
State Higher Educational Institution "National Mining University", Dnipropetrovs'k, Ukraine

ABSTRACT: In this paper, a model of the galvanic battery is presented. This model is composed of equations which are suitable for the simulation of various types of galvanic batteries if the proper coefficients are used. The model can be used as a tool for the calculation of electric vehicle performances, such as the estimation of fuel-saving potential and the verification of efficiency of the state-of-charge management strategies. The technique for an evaluation of the necessary equations' coefficients on the rated data of a battery is given.

1 INTRODUCTION

The battery is the core element of either a hybrid or pure electric vehicle. Other components like electric motors, electric power converters of various types and even internal combustion motors are believed to have reached their maximal possible efficiencies. Their further improvement is hardly possible in the near future. Moreover, with the current performances of energy storage devices, which are primarily galvanic batteries, improvement of other electric components is inexpedient. For a hybrid electric vehicle (HEV) of any topology, it is the battery that limits fuel-saving and, usually, vehicle's dynamics.

To prove this statement, we have to consider the sources of fuel saving and their potential. A hybrid electric vehicle derives its economy from three basic processes: recovery of vehicle's kinetic energy during braking, improvement of internal combustion engine (ICE) operation regime and turning the engine off during stops (microhybrid technologies).

Utilization or recovery of kinetic energy means capturing it during braking (otherwise it dissipates as heat on the brake disks) and storing it in the battery or other device like an ultra capacitor or a super fly wheel. This approach gives the best fuel saving for driving patterns with frequent starts and stops. Also, it can be done only if the electric part of the power train has enough power rating to provide necessary deceleration. In most modern HEVs only a small part of the kinetic energy is recovered, mainly due to constraints from the battery charging rate.

Shifting the ICE's operation point into more efficient zone is done mainly in vehicles of a series

topology. The engine rotates with constant speed and torque and charges the battery with optimal current. Then the ICE is turned off and the vehicle propels itself using the energy accumulated. This principle can also be realized in parallel hybrids, though it requires a more complicated control approach. Generally, the principle is best suited for driving patterns with relatively long cruising. Again, the battery energy capacity defines the vehicle's efficiency.

Finally, many modern vehicles employ so called microhybrid technologies. In microhybrids, there is no mechanical connection between the electric motor and the wheels and no electric power train. The electric part of the vehicle is used only to provide operation of auxiliary equipment when the ICE is turned off. This approach is especially effective for city driving with frequent stops at traffic lights, saving fuel and reducing CO_2 emissions.

The potential of the three sources is shown in Fig. `1.

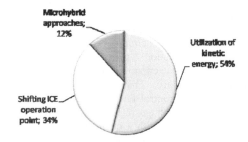

Figure 1. Potential of the main fuel-saving sources

The value of maximum fuel saving, provided by all three technologies, is taken to be 100%. So, the recovery of kinetic energy provides the biggest share of maximal possible fuel saving, but, at the same time, this principle requires the highest demands for battery ratings, particularly, its charging current.

Today, electric double-layer capacitors, or ultra capacitors, have emerged on the market and became rather cheap. Their application for energy recovery in HEVs is very promising, but the need to stabilize the voltage makes such solutions still very expensive. So, galvanic batteries will remain the prevailing energy storage in the near-term scenario.

All energy-saving strategies for electric vehicles rely on certain management of the battery's state of charge (SOC). To apply one, it is necessary to know the behavior of the battery and its limiting factors during charging/discharging modes. We shall try to provide researchers with the tool for the estimation of efficiencies of their energy-management strategies and verification of vehicles' performances.

2 BATTERY MODEL

Assumptions. Physical phenomena occurring in galvanic batteries are complicated and, generally, described by nonlinear equations. In context of calculations for HEVs the model of battery should be simplified as much as possible. Thus, the following assumptions should be taken:
- the battery's internal resistance is constant during charge and discharge cycles and does not depend on the amplitude of the current;
- all required coefficients in the equations and model parameters are taken from discharge characteristics, and accepted the same for charge mode;
- the temperature does not affect the model's behavior;
- the self-discharge of the battery is neglected;
- the memory effect is neglected.

General equation. The battery model represented below reflects all electrical processes occurring in galvanic elements of various types. Equations reflect the voltage dynamics under current variation and take into account the value of open circuit voltage (OCV) as a function of state of charge (SOC). The effect of polarization is induced into equations in order to improve the accuracy of simulations for no-load operation modes.

The battery voltage is described by the following equation:

$$Vbatt = E_0 - K\frac{Q}{Q-it} \cdot it - R \cdot i + A\,exp(-B \cdot it) - K\frac{Q}{Q-it} \cdot i^*$$

where $Vbatt$ – battery voltage, V;

E_0 – battery constant electromotive force, V;

K – polarization constant (V/(Ah)) or polarization resistance (Ω);

Q - battery capacity, Ah;

it – actual battery charge, Ah;

A – exponential zone amplitude, V;

B – exponential zone inverse time constant, Ah;

R - internal resistance, Ω;

i – battery current, A;

i^* – filtered current, A.

Model limitations:
- minimum voltage of non-load battery is 0 V, maximum voltage is $2 \times E_0$;
- minimum battery capacity is 0 Ah, maximum capacity is Q.

The equation contains the component of voltage drop caused by the filtered current flowing through polarization resistance. This element of equation reflects the process of slow voltage decrease down from its maximum level with the rate proportional to the current magnitude. Using filtered current allows one to exclude algebraic feedbacks, a typical problem of galvanic batteries simulation.

Another problem is that the OCV depends nonlinearly from SOC. This effect can be simulated using so called polarization voltage. The last elementin the equation represents nonlinear function of battery voltage from discharge current and actual battery charge.

The exponential discharge zone in the equation is typical for Li-Ion batteries. Batteries of other types (Lead-Acid, NiMH and NiCD) have hysteresis between charge and discharge, which does not depend on their actual SOC. The voltage drop caused by this phenomenon can be calculated using nonlinear dynamical system given below:

$$Exp(t)' = B \cdot |i(t)| \cdot (-Exp(t) + A \cdot u(t))$$

where $Exp(t)$ – voltage in the exponential zone, V;

$i(t)$ – battery current, A;

$u(t)$ – charge or discharge mode.

Voltage in the exponential zone depends on its initial value $Exp(t_0)$ and operation mode (charge or discharge). It should be taken into account that charge and discharge processes' description depends on the type of battery. Proper equations for different batteries are given in the table below.

Table 1. Functions of voltage drop for various types of batteries.

Type of battery	Equations for charge and discharge modes
Lead-Acid	Discharge $$Vbatt = E_0 - R \cdot i - K \frac{Q}{Q-it} \cdot (it + i^*) + Exp(t)$$ Charge $$Vbatt = E_0 - R \cdot i - K \frac{Q}{it - 0.1 \cdot Q} \cdot i^* - K \frac{Q}{Q-it} \cdot it + Exp(t)$$
Li-Ion	Discharge $$Vbatt = E_0 - R \cdot i - K \frac{Q}{Q-it} \cdot (it + i^*) + A\,exp(-B \cdot it)$$ Charge $$Vbatt = E_0 - R \cdot i - K \frac{Q}{it - 0.1 \cdot Q} \cdot i^* - K \frac{Q}{Q-it} \cdot it + A\,exp(-B \cdot it)$$
NiMH and NiCd	Discharge $$Vbatt = E_0 - R \cdot i - K \frac{Q}{Q-it} \cdot (it + i^*) + Exp(t)$$ Charge $$Vbatt = E_0 - R \cdot i - K \frac{Q}{|it| - 0.1 \cdot Q} \cdot i^* - K \frac{Q}{Q-it} \cdot it + Exp(t)$$

Having these equations and interdependencies between voltage components and battery current, we can compose the universal model of galvanic battery (see Fig. 2).

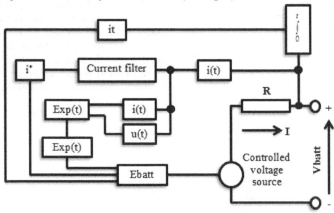

Figure 2. Battery model

Determination of coefficients in the equation. To determine the equation coefficients and model parameters it is necessary to measure three points at the discharge curve: the fully charged voltage (*Vfull*), the end of the exponential zone (voltage and charge) and the end of the nominal zone (voltage and charge), as shown in Fig. 3

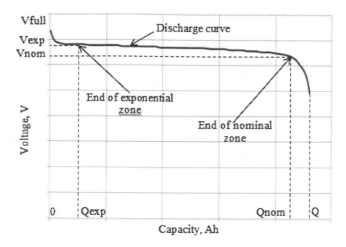

Figure 3. Typical discharge curve

Exponential part of equations is calculated according to the expressions

$$A = V_{full} - V_{exp},$$ (3)

$$B = \frac{3}{Q_{exp}}.$$ (4)

Coefficient K is calculated by:

$$K = \frac{\left[V_{full} - V_{nom} + A(exp(-B \cdot Q_{nom}) - 1)\right] \cdot (Q - Q_{nom})}{Q_{nom}}$$ (5)

Simulation. A typical Lithium-Ion battery was simulated using equations (1-5) and the structure given in Fig. 2. The dependence of battery voltage on capacity – simulated curve and experimental one - are given below.

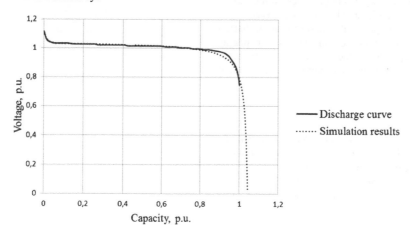

Figure 4. Simulation results

As we can see, the model rather accurately reflects the discharge process. Dynamical behavior of the battery during several charge/discharge cycles is shown in the figures below. The curves are built in per units.

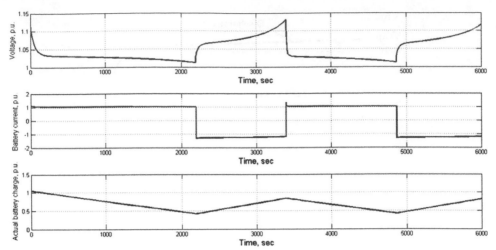

Figure 4. Dynamic processes during charge/discharge cycles

3 CONCLUSIONS AND FUTHER STUDIES

The structure and dependencies reflect correlations between battery state of charge, its voltage and current in various operation modes. Thus the model can be used for simulation of energy exchange processes in hybrid or pure electric vehicles, estimation of necessary battery parameters in order to obtain target performances on fuel saving or vehicle's dynamics and for other purposes.

In order to improve the description of battery behavior it is necessary to consider the dependence of battery capacity on the magnitude of discharge current. It is especially important when dealing with Lead-Acid batteries.

REFERENCES

Beshta O. *Design of electromechanical system for parallel hybrid electric vehicle.* Proceedings of the European Control Conference.
Benn H. 2008. *Simulations of batteries and their physical effects.* IEEE trans. issue, Vol.1: 123-125.

Power Engineering, Control and Information Technologies in Geotechnical Systems – Pivnyak, Beshta
& Alekseyev (eds)
© 2015 Taylor & Francis Group, London, ISBN 978-1-138-02804-3

Binarization algorithm of rock photo images on inhomogeneous background

P. Pilov, M. Alekseev & I. Udovik
State Higher Educational Institution "National Mining University", Dnipropetrovs'k, Ukraine

ABSTRACT: This work is dedicated to description of modified binarization algorithm of rock color photographs with the purpose of object selection that are different in size, minerals fraction, represented on inhomogeneous background for further possibility of automatization and grain size analysis of their content. Results of experimental research of proposed algorithm possibilities that prove the increase of binarization authenticity and minerals fraction contour selection in comparison with initial method are represented.

1 INTRODUCTION

One of the main stages of technological process in mining industry is grain size determination. It represents calculation of percent weight content of rock of different in size fraction (Pilov 2013). Sieve method, based on rock separation through the set of different in size units is traditionally applies for this issue solution. This method requires essential application of hand work and time spending that embraces it's usage in some cases, for example, rapid analysis realization. Therefore, development of new technologies allows to realize rock grain size determination in automatized or automatic modes with minimal expenses of handwork and sufficient level of accuracy is actual task.

One of possibilities for task solution of grain size composition express-testing conduction is application of photo images that were received during coal extraction or rock processing (Kornienko 2006). While using such approach one of the main stages that determines accuracy of eventual result is the background processing of initial photograph (Kornienko & Gluhov 1990). This stage includes such operations as noise filtering, contrast enhancement, binarization, etc. The way of performance essentially depends upon the quality and characteristics of initial images. Final purpose of processing is receiving the data in a suitable look for automatic (automatized) grain size composition.

The main issues that complicate background processing of images are inhomogeneity of background and lightening, presence of degraded and low-contrast areas, "cohesion" and overlapping (apposition) at each other.

For instance, inhomogeneity of background and lightening embraces application of traditional methods of filtration (median filter or Wiener filter (Pratt 2001)) and improving of objects division on the basis of histogram method application or transformation of contrast stretching (Gonzalez & Woods 2006). Presence of low-contrast and light-struck areas makes it impossible to detect an immediate contour of objects of interest on basis of liminal (Huang and Vang method (Chi, Yan & Pham 1998)) or gradient (Sobel operator (Pratt 2001; Gonzalez & Woods 2006)) methods.

2 FORMULATION OF THE PROBLEM

In work (Pilov, Akhmetshina, Yegorov & Udovick 2014) proposed algorithm of photographs of rock binarization that allows to emphasize contours of objects of interest with the purpose of further grain size analysis conduction. Its disadvantage is enhancement of considerable amount of non-informative (false) contours that reflecting structure, not object of interest borders, and also insufficient sensitivity during separation of contacting objects with same brightness properties that determined by usage of only one (red) color channel.

Modified algorithm of binarization of rock color photographs that allows increasing authenticity of their analysis by means of all color channels is proposed.

3 MATERIALS FOR RESEARCH

Proposed algorithm consists of following steps:

1) increasing of brightness and contrast of red color channel I_r^1 (chose as more informative based upon experimental research) after scaling on a bit [0,1] ;

2) removal of inhomogeneous background (it should be emphasized that its choice has a great impact on photo processing method);

3) brightness transformation of received image with the purpose of objects separation improvement;

4) contour enhancement;

5) binarization of received image;

6) synthesis of image that based on indices of maximal value of color partial of each pixel;

7) suppression of little marks differences in formed image;

8) execution of steps 3 and 4;

9) step 5 for image that was received after execution of previous stage;

10) formation of total image on the basis of two binary photographs combination that were received by steps 6 and 7.

Increasing of brightness and contrast for image I_r^1 is realized by formula:

$$I_{x,y}^2 = \left(\left(I_r^1 \right)_{x,y} \right)^{\left(1 - \left(I_r^1 \right)_{x,y}\right)^{-\left(I_r^1 \right)_{x,y}}} \tag{1}$$

for all image pixels that brightness satisfies to the condition $\left(I_r^1 \right)_{x,y} \leq \overline{I_r^1}'$, где $\overline{I_r^1}' = \left(\overline{I_r^1} + 0.5 \right) / 2$, but $\overline{I_r^1}$ – average over image $\overline{I_r^1}$

Transformation of each image's background pixel brightness of I^2 is realized by the following algorithm.

1. Image I^2 splits on interlocking windows with sizes 3x3 pixels (in all windows transformations for proposed algorithm apply windows of such size), and for all pixel $w_{x,y}$ of current window realizes transformation:

$$w_{x,y} = w_{x,y} + \frac{\overline{w} - w_{x,y}}{2} \tag{2}$$

where $\overline{w}$ – its average brightness, as a result of which the image forms I^3 .

2. Receiving of image I^4 on the basis of power-mode transformations that provide nonlinearity of brightness transformation:

$$I_{x,y}^4 = \left(I_{x,y}^3 \right)^{\left(1 - \left(1 - I_{x,y}^3\right)^{I_{x,y}^3}\right)} \tag{3}$$

This step is one of the main during background suppression (reducing t white color).

3. Successive two-time appliance of the rank filtration (to the maximum for area 3x3 pixel) (Shliht 1997) with following execution of 1st step, as a result of which forms image I^5 .

4. Receiving of image I^6 by following formula:

$$I_{x,y}^6 = \left(I_{x,y}^5 \right)^{\left(1 - I_{x,y}^5\right)} \tag{4}$$

5. Formation of image I^7 in a certain way:

$$I_{x,y}^7 = \left(I_{x,y}^6 \right)^{\left(1 + \left(I_{x,y}^6 \right)^{1 - I_{x,y}^6}\right)} \tag{5}$$

6. On the final step the sequential replacement of all brightness values occurs that exceeds $\overline{M}$, on 1. This value calculates as average on brightness that satisfies condition:

$$T_1 < I_{x,y}^7 < 1 \tag{6}$$

Such process happens until realizes condition $\overline{M} > T_1 + (1 - T_1)/3$. It is recommended to use $T_1 = 0.9$ on the basis of experimental research. Then the rank filtration is applied (on the maximum) as a result of which image I^8 forms. Transformation of brightness with the purpose of objects separation improving by making use of following formulas:

$$\begin{cases} I_{x,y}^9 = \left(I_{x,y}^8 \right)^{\left(1 - \left(I_{x,y}^8 \right)^{1 - I_{x,y}^8}\right)} & ; I_{x,y}^8 \leq \overline{I^8} \\ I_{x,y}^9 = \left(I_{x,y}^8 \right)^{\left(1 + \left(I_{x,y}^8 \right)^{I_{x,y}^8}\right)} & ; I_{x,y}^6 > \overline{I^8}, \overline{I^8} \leq 0.5 \\ I_{x,y}^9 = \left(I_{x,y}^8 \right)^{\left(1 - \left(I_{x,y}^8 \right)^{I_{x,y}^8}\right)} & ; I_{x,y}^6 > \overline{I^8}, \overline{I^8} > 0.5 \end{cases} \tag{7}$$

$$I_{x,y}^{10} = \left(I_{x,y}^9 \right)^{\left(1 + \left(I_{x,y}^9 \right)^{1 + I_{x,y}^9}\right)} \tag{8}$$

where $\overline{I^8}$ – average value of brightness of image I^8 pixels, which intensity is less than 1. After that realizes smoothing on the basis of formula (2), as a result of which forms image I^{11} .

Gaussian's Laplasian (Gonzalez & Woods 2006) is used for contour enhancement with value $\sigma = 0.5$, as a result of which forms mask with sizes 3x3 pixel. After application of received mask to photograph I^{11} forms image I^{12}, for which calculates minimal value $I^{12}_{\min}$. As one of the main disadvantages of Gaussian's laplasian application is formation of double contours, so for all image pixels $I^{12}_{\min}$, which brightness values are reset to zero. Then execute transformations that intensify differences between contour color and background:

$$I^{13}_{x,y} = \left(I^{12}_{x,y}\right)^{\left(\left(I^{12}_{x,y}\right)^{1+I^{12}_{\min}} - I^{12}_{\min}\right)^{1+I^{12}_{\min}}} \tag{9}$$

$$I^{14}_{x,y} = \left(I^{13}_{x,y}\right)^{1-I^{13}_{x,y}} \tag{10}$$

Binarization of received image I^{14} realizes with use of condition $I^{14}_{x,y} \geq \overline{I^{14}}$, where $\overline{I^{14}}$ – average value by photograph I^{14} pixels, intensified values of what is more than 0. Than executes rank filtration order 7, as a result of which forms first binary image I^{1}_{bin}.

It is necessary to use information about all color channels of initial image for objects separation improving with different brightness and color characteristics. In proposed algorithm it realizes in a certain way: image I^{1}_{ind} forms on the basis of indices of maximal value of color partial of each pixel and scaling on a bit $[0,1]$.

After that suppression of little mark differences between brightness values of same pixel in different color channels is realized. It materialized by means of stamping brightness of those pixels of image I^{1}_{ind}, which difference of intensities in any two channels does not exceed values $\overline{d_{ch}}/4$. At this $\overline{d_{ch}}$ – average value of differences between intensities of chosen pair of color channels. Then medial filtration of received photograph occurs and the rank filtration realizes twice (on the maximum).

Received image is exposed to transformation by formula (2), as a result of which forms photograph I^{2}_{ind}. Transformation of image I^{2}_{ind} brightness with the purpose of objects separation improving is materialized by formula:

$$\left(I^{3}_{ind}\right)_{x,y} = \left(\left(I^{2}_{ind}\right)_{x,y}\right)^{1-\left(\left(I^{2}_{ind}\right)_{x,y}\right)^{-\left(I^{2}_{ind}\right)_{x,y}}} \tag{11}$$

After the rank filtration (on the maximum) for area 3x3 pixel is realized twice.

Contour enhancement in image I^{3}_{ind} also realizes on the basis of Gaussian's laplasian application with is described by the above-mentioned parameters. After application of received mask to photograph I^{3}_{ind} the image I^{4}_{ind} forms, for which the minimal value $I^{4}_{\min}$ is calculated, and for pixels which brightness values are reset to zero. Then intensification of differences between contour and background on the basis of transformation is materialized:

$$\left(I^{5}_{ind}\right)_{x,y} = \left(\left(I^{4}_{ind}\right)_{x,y}\right)^{1-\left(\left(I^{4}_{ind}\right)_{x,y}\right)^{1+I^{4}_{\min}} - I^{4}_{\min}/2} \tag{12}$$

and formula (10) as a result of which forms image I^{6}_{ind}.

Binarization of this image is materialized also for the photograph I^{14}, as a result of which the image I^{2}_{bin} is formed.

Total image I^{out} is the result of photographs I^{1}_{bin} and I^{2}_{bin} interflowing (on the basis of brightness maximum) and further appliance of rank filtration of order 4.

Experimental results. Represented modified algorithm was used for images processing that contains different rock minerals on inhomogeneous background. Great example, the photograph, color channels of which are represented on Fig. 1 a – 1 c.

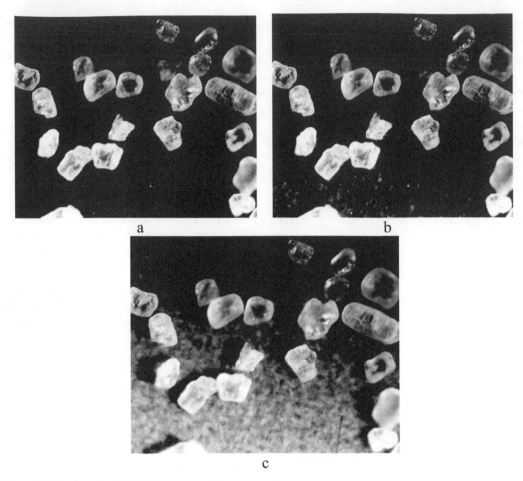

Figure 1. Color channels of initial image: a – red; b – green; c – blue

On Fig. 2 represented result of increasing brightness and contrast of red color channel as a result of which, in particular, observing improving of visual legibility of stone in left upper corner. On Fig. 2 b results of background rejection (substitution by white color) are given, but on a Fig. 2 c – improving of objects separation. Final stage provides also decreasing non-informative detalization of objects parts that caused by specialties of their form and structure. On Fig. 2 d reflected result of objects separation improving on image that was received on the basis of color channel indices with maximal brightness for each pixel. This image allows separating different colors better.

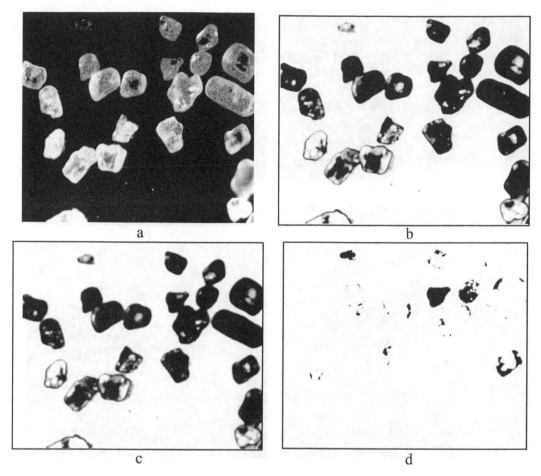

Figure 2. Results of initial image processing: a – increasing of brightness and contrast for red color channel; b – background rejection; c – objects separation improving (image I^{11}); d – objects separation improving (изображение image I^3_{ind})

On Fig. 3 represented results of red color channel binarization for initial algorithm. On Fig. 3 b – 3 d demonstrated application of modified binarization algorithm. At this on Fig. 3 b educed image I^1_{bin} (red color channel binarization), on Fig. 3 c – photograph I^2_{bin}, on Fig. 3 d – final result. It is evident that application of information about color channels allowed improving separation of objects with different colors. By comparison of red color channel binarization of initial (Fig. 3 a) and final (Fig. 3 b) results of modified algorithm we can see that in last case quantity of false contours decreases and also improves objects separation.

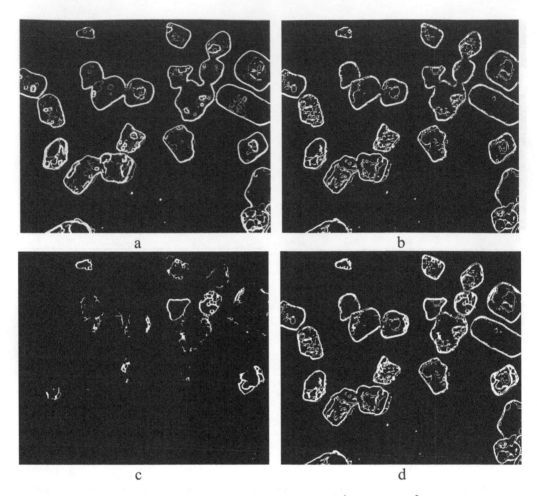

Figure 3. Results of binarization: a – initial algorithm application; b – image I_{bin}^{1} ; c – image I_{bin}^{2} ; d – final image

4 CONCLUSIONS

Proposed modified algorithm of binarization of rock color photographs with the purpose of objects of interest structuring for further grain-size analysis conduction allows to increase authenticity of their detecting in comparison with initial method. The main disadvantages of proposed approach are false contours detecting that is caused by objects structure, impossibility of light-struck objects processing (that stated in interpretation of few such objects as single structure) and high computational complexity.

REFERENCES

Pilov P.I. 2013. *The description of the specific surface of the reduction products based on the dispersion distribution function.* Benefication of the minerals, Issue 53(94): 60-68.

Kozin V.Z. 2005. *The control of the benefication processes* (in Russian). Yecaterinburg: HS: 303.

Kornienko V.I. 2006. *The logical algorithms of the binary image processing in the optical granulometr of the crushed materials* (in Russian). Dnipropetrovs'k:Science Bulletin of the National Mining University, Issue 11: 89-90.

Kornienko V.I., Gluhov V.V. 1990. *The problems of the mathematic support development for the optical granulometr of the crushed materials* (in Russian). Metal and mining manufacture, Vol. 2: 68-71.

Pratt W.K. 2001. *Digital Image Processing.* New York; Chichester; Weinheim; Brisbane: John Wiley and Sons Inc.: 723.

Gonzalez R., Woods R. 2006. *Digital image processing [transl. from English, edited by*

Chochia P.A.] (in Russian). Moscow: Technosphera: 1070.

Jähne B. 2006. *Digital image processing [transl. from English, edited by Izmaylovoy A.M.]* (in Russian). Moscow: Technosphera: 1070.

Chi Z., Yan H., Pham T. 1998. *Fuzzy algorithms: With Applications to Image Processing and Pattern Recognition.* Singapore; New Jersey; London; Hong Kong: Word Scientific: 225.

Pilov P.I., Akhmetshina L.G., Yegorov A.A., Udovick I.M. 2014. *The Transformation of Rock Photos with Heterogeneous Backgroundground for the Granulometric Composition Check* (in Russian). Kherson: Bulletin of the Kherson National Technical University, Issue 3 (54): 50 – 54.

Shliht G.Y. 1997. *The digital processing of the color images* (in Russian). Moscow: Ecom: 336.

Compensation of the cogging torque by means of control system for transverse flux motor

E. Nolle
Esslingen University of Applied Sciences , Esslingen, Germany

O. Beshta & M. Kuvaiev
State Higher Educational Institution "National Mining University", Dnipropetrovs'k, Ukraine

ABSTRACT: The problem of the cogging torque influence of gearless drives with transverse flux motor is described. The way of the cogging torque compensation for transverse flux motor has been proposed. The task of the delay between current reference and the reproduced electromagnetic torque and a ratio correction factor, which depend on the frequency, has been solved. Recommendations are given on using the methods of control law synthesis.

1 INTRODUCTION

Gearless drives with synchronous motors with permanent magnets (PMSM) are used in more and more mechanical applications that require high torque, because of the importance of technology improvement by means of electric drives (O. Beshta 2012). One of the alternatives to replace the standard PMSM in gearless drives is a transverse flux motor (TFM) with permanent magnets. The magnetic flux of TFM runs transversely to the plane of rotation of the rotor. The magnetic field still rotates because of the geometry of the motor. There are geometrical shifts between phases by 120° electrical degrees. TFM has advantages such as high torque at low speed because it can have a big number of poles, small loss compared to the same conventional PMSM because the TFM has no front part in its winding.

2 RESEARCH

Along with the advantages TFM has some of disadvantages: high leakage inductance and torque ripples. These ripples have a negative influence on the accuracy of regulation and bring acoustic noise and vibration, which often is not acceptable in gearless drive applications. Much of the ripple of the torque happen because of cogging torque. Particularly significant influence has cogging torque in low speed gearless drives, because of large magnetic induction of permanent magnets in air gap. There are many ways to reduce the cogging torque for classical PMSM by constructional changes, such as bevel of stator teeth or permanent magnets on the rotor, choosing the optimal combination of number of teeth of the stator and rotor poles, improving the system of the excitation on the rotor and stator design improvements. These methods are also suitable for TFM (Wan-Tsun Tseng 2008), but they lead to increased cost and complexity of electric motor production. Another way is creating such a harmonic component of stator current that would compensate cogging torque.

Cogging torque is a function of rotor position. It is difficult to calculate the value of cogging torque analytically, because it is a function of the angle of the rotor position and the result will not be too exact. It is possible to remove the torque function of the angle of the rotor position using special equipment at low speed and without load, but this method requires substantial financial costs. An alternative is to calculate the cogging torque with a specialized program for the calculation of electromagnetic fields. One of these specialized programs is Maxwell (ANSYS), which is a finite element design package.

Maxwell calculates a cogging torque for TFM in idling mode of a generator. The rotational speed is set at 1°rad/sec, the required simulation time is calculated very easy. The TFM with external rotor and the following parameters was taken for example:

$$T_n = 100Nm \text{, } n_n = 187,5\,\frac{1}{\min}\text{, } f_n = 50Hz \text{,}$$

$$I_n = 4,5A \text{.}$$

In Fig. 1 the cogging torque function of the angle of rotation of the rotor, which is obtained in the program Maxwell, is shown.

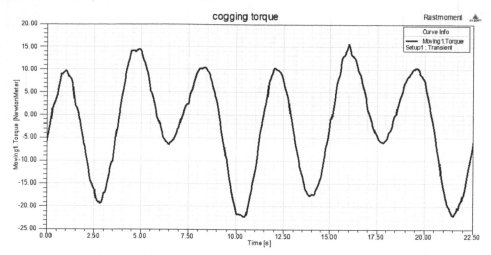

Figure 1. Cogging torque of TFM simulated in Maxwell

The equation for cogging torque function of the rotor was received with the help of Maxwell special tools. The harmonic analysis of cogging torque was done for this purpose.

As a result of this analysis we've got amplitude (Fig. 2) and phase shifts (Fig. 3) of harmonics for the cogging torque.

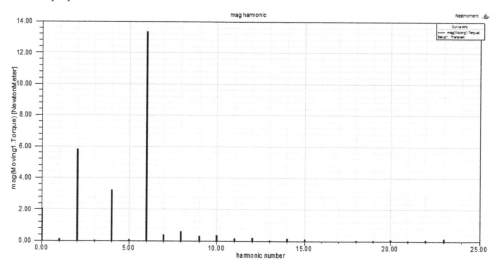

Figure 2. Amplitude of harmonics of the cogging torque calculated in Maxwell

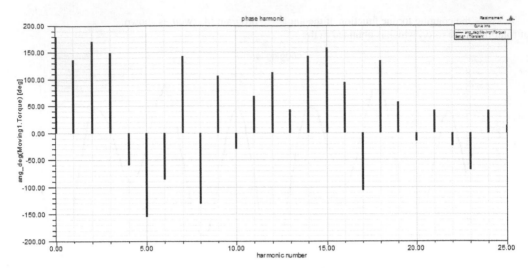

Figure 3. Phase shifts of harmonics of the cogging torque calculated in Maxwell

Using the results of harmonic analysis we've got the equation for cogging torque as a function of the rotor angle in the form of a Fourier series:

$$
\begin{aligned}
T_{cog} &= T_{cog\,0} + T_{cog\,1} \times \\
&\times \cos(\theta + \psi_1) + T_{cog\,2} \times \\
&\times \cos(2 \cdot \theta + \psi_2) + \dots + \\
&+ T_{cogk} \cdot \cos(k \cdot \theta + \psi_k)
\end{aligned}
\tag{1}
$$

According to equation the function $T_{cog} = f(\theta)$ was generated in Simulink application (Fig.4).

Tcog[Nm]

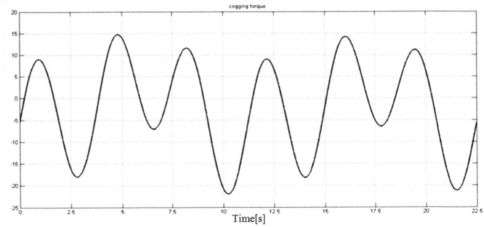

Time[s]

Figure 4. Cogging torque of TFM in Simulink

To compensate the cogging torque, the classical vector control system with the rotor position sensor (digital position encoder) was taken because the value of the cogging torque depends on the rotor angle.

In the classical vector control system for PMSM the component of current i_d is always zero. At the input of the i_q current regulator the appropriate reference for harmonic components should be applied so that they could reproduce the electromagnetic torque, which would be in opposite phase to the cogging torque:

$$i_{cogq} = i_{cogq\ 1} \cdot \cos(\ \theta + \psi_1 + \pi\ ) +$$
$$+\ i_{cogq\ 2} \cdot \cos(\ 2 \cdot \theta + \psi_2 + \pi\ ) +$$
$$+ \ldots + i_{cogk} \cdot \cos(\ k \cdot \theta + \psi_k + \pi\ ) = \qquad (2)$$
$$= -(i_{cogq\ 1} \cdot \cos(\ \theta + \psi_1\ ) +$$
$$+\ i_{cogq\ 2} \cdot \cos(\ 2 \cdot \theta + \psi_2\ ) +$$
$$+ \ldots + i_{cogk} \cdot \cos(\ k \cdot \theta + \psi_k\ ))$$

The T_{cog0} must not be extra compensate, because of this harmonic is constant. There is a delay between reference current and the reproduced electromagnetic torque. In addition to delay there is a ratio correction factor that changes as a function of frequency. To find these parameters it is necessary to make a transfer function of the current controller and the motor current. Than amplitude-frequency characteristic and phase-frequency variation for this transfer function have been calculated (Fig. 5):

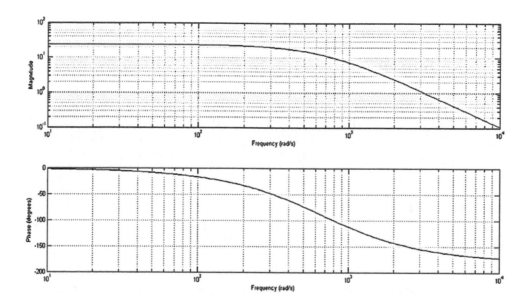

Figure 5. Amplitude-frequency characteristic and phase-frequency variation

With new parameters the equation (2) changes to:

$$i_{cogq} = -(i_{cogq\ 1} \cdot k_{a1}(f) \times$$
$$\times \cos(\ \theta + \psi_1 + \gamma_1(f)\ ) +$$
$$+\ i_{cogq\ 2} \cdot k_{a2}(f) \times \qquad (3)$$
$$\times \cos(\ 2 \cdot \theta + \psi_2 + \gamma_2(f)\ ) +$$
$$+ \ldots + i_{cogk} \cdot k_{ak}(f) \cdot$$
$$\cdot \cos(\ k \cdot \theta + \psi_k + \gamma_k(f)\ )$$

where $\gamma_{1\ldots k}(f)$ – delay function, $k_{a1\ldots ak}(f)$ – ratio correction factor function of function, f – frequency.

This method was tested for a speed of 12,6rpm.

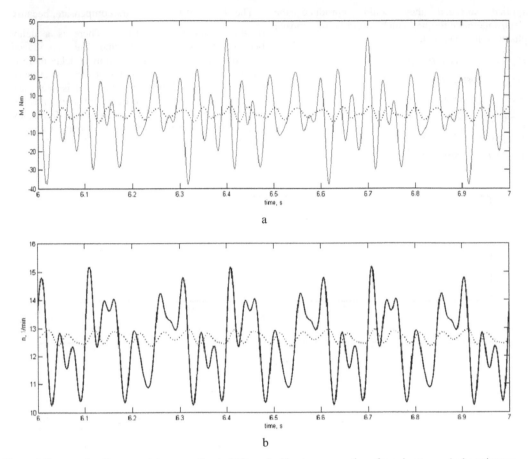

Figure 6. Comparative diagrams a) torque on the shaft b)speed without compensation of cogging torque (red continuous line) and with compensation of cogging torque (blue interrupted line)

As the diagram (Fig. 6) shows, the amplitude of fluctuation of speed (Fig. 6,b) without compensation of cogging torque is about 19%, and with compensation is only 2%, ripples of torque on the shaft (Fig. 6,a) with the compensation are only 1/10 of the torque ripples on the shaft without compensation.

3 CONCLUSIONS

The method of counter-phase current provides high-quality and cheap way to compensate the cogging torque for the TFM, that has a negative influence on the quality of the production process, especially in gearless drives at low speed. Compensation of torque ripples can be done by the vector control system with the rotor position sensor (digital position encoder) because the cogging torque is the function of the rotor angle. The delay between current reference and the reproduced electromagnetic torque and a ratio correction factor must change depending on the frequency. Determination of amplitude-frequency characteristic and phase-frequency variation, as the generalized function of the motors parameters is the object for a following research.

REFERENCES

Beshta O. 2012. *The frequency-controlled electric drives application in tasks of higher power-efficient of technological processes* (in Russian). Dnipropetrovs'k: Scientific bulletin of the National Mining University, Issue 4: 98-107.

Tseng Wan-Tsun 2008. *Theoretische und experimentelle Untersuchungen zu einem permanentmagneterregten Transversalfluss-Synchronlinearmotor in Sonderbauform.* Dissertation: 28-31.

Power Engineering, Control and Information Technologies in Geotechnical Systems – Pivnyak, Beshta & Alekseyev (eds)
© 2015 Taylor & Francis Group, London, ISBN 978-1-138-02804-3

Control of tandem-type two-wheel vehicle at various notion modes along spatial curved lay of line

O. Beshta, V. Kravets, K. Bas, T. Kravets & L. Tokar
State Higher Educational Institution "National Mining University", Dnipropetrovs'k, Ukraine

ABSTRACT: Wheeled vehicle is considered as a material point under the conditions of non-uniform movement along curved spatial lay of line. Hodograph in a class of spiral lines describes kinematics of a vehicle. A kinetostatics problem of tandem-type two-wheel vehicle is being solved. Equivalent contact dynamics are being determined.

1 INTRODUCTION

A problem of dynamic design in the context of two-wheel vehicle controllability as well as dynamic burden of its design and road surface is important in terms of various motion modes (accelerated, decelerated, and steady) along curved spatial lay of line in junctions, at various gradients, on straight and turns as well as within other curved areas (Khachaturov 1976; Martunjuk, Lobas & Nikitina 1981). The problem solution will help determine equivalent contact control force making analysis of the required control facilities.

2 FORMULATION THE PROBLEM

Hodograph of vehicle motion along curved spatial lay of line is supposed as given. A vehicle is considered as material point of known mass moving under the gravity, given aerodynamic forces, and searched equivalent contact moving (control) forces – resulting reaction of contact with reference trajectory (lay of line). Curved spatial reference trajectory that is lay of line is identified by hodograph in motionless earth reference.

3 MATERIALS FOR RESEARCH

Hodograph correspondent to real trajectory of vehicle motion (Martunjuk, Lobas & Nikitina 1981) sought in a class of spiral lines (Kravets 2013) is specified in motionless (earth) reference as follows

$$\bar{r}(t) = \|\rho_0 \rho_1 \rho_2 \rho_3\| \begin{Vmatrix} 1 \\ t \\ t^2 \\ t^3 \end{Vmatrix} (\bar{i}\cos\omega t + \bar{j}\sin\omega t) + \bar{k} \|h_0 h_1 h_2 h_3\| \begin{Vmatrix} 1 \\ t \\ t^2 \\ t^3 \end{Vmatrix}$$

where $\rho_i h_i (i = 0,1,2,3)$ are varied parameters determined on specified boundary conditions; ω is mean angular turn velocity equal to $\omega = \dfrac{\varphi_0}{t_0}$. Here φ_0 is complete turn angle; and t_0 is required time of turn passing.

Following figure demonstrates lay of line as a trajectory of wheeled vehicle motion in terms of curved area being adequate to proposed hodograph:

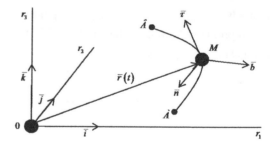

Figure 1.

Here $\bar{i}, \bar{j}, \bar{k}$ are orts of earth (motionless) reference; and $\bar{\tau}, \bar{n}, \bar{b}$ are orts of movable, natural axes.

It is obvious that hodograph is represented in a well-known representation form (Lobas 2009):

$$\bar{r}(t) = \bar{i}\, r_1 + \bar{j}\, r_2 + \bar{k}\, r_3 .$$

Here vector components are assumed as:

$$r_1 = \|\rho_0\rho_1\rho_2\rho_3\| \left\|\begin{matrix} 1 \\ t \\ t^2 \\ t^3 \end{matrix}\right\| \cos\omega t,$$

$$r_2 = \|\rho_0\rho_1\rho_2\rho_3\| \left\|\begin{matrix} 1 \\ t \\ t^2 \\ t^3 \end{matrix}\right\| \sin\omega t, \quad r_3 = \|h_0h_1h_2h_3\| \left\|\begin{matrix} 1 \\ t \\ t^2 \\ t^3 \end{matrix}\right\|.$$

Here you can find hodograph of a vehicle motion:
1. steady $(V_{1A} = V_{1B})$ motion within horizontal $(h_j = 0; \ j = 0,1,2,3)$ straight $(\omega = 0)$ lay of line:

$$\overline{r}(t) = \overline{i}\left(r_{1A} + V_{1A} \cdot t\right);$$

2. unsteady: $(V_{1A} < V_{1B})$ – accelerated; $(V_{1A} > V_{1B})$ – decelerated motion within horizontal $(h_j = 0; \ j = 0,1,2,3)$ straight $(\omega = 0)$ lay of line:

$$\overline{r}(t) = \overline{i}\left(r_{1A} + V_{1A} \cdot t + \frac{1}{4}\cdot\frac{V_{1B}^2 - V_{1A}^2}{r_{1B} - r_{1A}}t^2\right);$$

3. steady $(V_{1A} = V_{1B}, V_{3A} = V_{3B} = 0)$ motion within profile-inclined lay of line if $(r_{3A} < r_{3B})$ – rise and $(r_{3A} > r_{3B})$ – incline:

$$\overline{r}(t) = \overline{i}\left(r_{1A} + V_{1A} \cdot t\right) +$$
$$+ \overline{k}\left(3 - 2\frac{V_{1A}}{r_{1B} - r_{1A}}\cdot t\right)\left(\frac{V_{1A}}{r_{1B} - r_{1A}}\right)^2 r_{3B}\cdot t^2.$$

Here using Cartesian coordinate system lay of line profile is represented in the form of square and cubic parabolas: $z = 3x^2 - 2x^3$, where

$$x = \frac{r_1(t) - r_{1A}}{r_{1B} - r_{1A}}, \quad z = \frac{r_3(t)}{r_{3B}}.$$

4. Unsteady motion within horizontal plane where direct- angle turn takes place:

$$\overline{r}(t) = \left[r_{1A} + \frac{12}{\pi^2}\left(\frac{V_{2A}}{r_{1A}}\right)^2(r_{2B} - r_{1A})t^2 - \frac{16}{\pi^3}\left(\frac{V_{2A}}{r_{1A}}\right)^3(r_{2B} - r_{1A})t^3\right] \times$$
$$\times\left(\overline{i}\cos\frac{V_{2A}}{r_{1A}}t + \overline{j}\sin\frac{V_{2A}}{r_{1A}}t\right)$$

Here lay of line plan in polar coordinate system is represented by square and cubic Archimedean spirals:

$$\frac{r(\varphi) - r_{1A}}{r_{2B} - r_{1A}} = 3\left(\frac{\varphi}{\pi/2}\right)^2 - 2\left(\frac{\varphi}{\pi/2}\right)^3,$$

where polar angle is: $\varphi = \omega t$; and
polar radius is: $r(\varphi) = r_1^2 + r_2^2$.

Moreover, when $V_{2A} = V_{1B}$ or $r_{1A} = r_{2B}$, it follows that: $r(\varphi) = r_{1A}$ at any φ, i.e. we obtain lay of line in the form of radial arc within the given interval: $0 \leq \varphi \leq \frac{\pi}{2}$.

Kinematics. Vector of vehicle linear velocity in the form of material point is determined on the specified hodograph as:

$$\overline{V} = \frac{d\overline{r}}{dt} \quad \text{or} \quad \overline{V} = \overline{i}\,\dot{r}_1 + \overline{j}\,\dot{r}_2 + \overline{k}\,\dot{r}_3.$$

Velocity value is determined with the help of scalar product:

$$\overline{V}\cdot\overline{V} = \upsilon^2 \quad \text{or} \quad \upsilon^2 = \dot{r}_1^2 + \dot{r}_2^2 + \dot{r}_3^2.$$

By definition, velocity value is also determined as a time derivative from the path:

$$\upsilon = \frac{ds}{dt} \quad \text{or} \quad \upsilon = \dot{S}.$$

Then the path of a vehicle within random time period is calculated by means of definite integral with variable upper limit:

$$S(t) = \int_0^t \upsilon(t)dt \quad \text{or} \quad S(t) = \int_0^t \sqrt{\dot{r}_1^2 + \dot{r}_2^2 + \dot{r}_3^2}\ dt.$$

When path is introduced as intermediate argument, velocity vector is represented as:

$$\overline{V} = \frac{d\overline{r}}{ds}\frac{ds}{dt} \quad \text{or} \quad \overline{V} = \frac{d\overline{r}}{ds}\dot{S}$$

And taking into account that: $\dfrac{d\overline{r}}{ds} = \overline{\tau}$, we obtain (Lobas 2009):

$$\overline{V} = \overline{\tau}\,\dot{S}.$$

It is obvious that velocity vector projection on tangent line ort to spatial trajectory is $\overline{\tau}\cdot\overline{V} = \overline{\tau}\cdot\overline{\tau}\,\dot{S}$. Determine velocity value as

28

follows: $V_\tau = \dot{S}$; to principal normal ort: $\bar{n} \cdot \bar{V} = \bar{n} \cdot \bar{\tau} \dot{S}$, $V_n = 0$; to binormal ort: $\bar{b} \cdot \bar{V} = \bar{b} \cdot \bar{\tau} \dot{s}$, $V_b = 0$. In terms of vector and matrix form we obtain:

$$\|\bar{\tau}\ \bar{n}\ \bar{b}\| \left\| \begin{matrix} V_\tau \\ V_n \\ V_b \end{matrix} \right\| = \|\bar{i}\ \bar{j}\ \bar{k}\| \left\| \begin{matrix} \dot{r}_1 \\ \dot{r}_2 \\ \dot{r}_3 \end{matrix} \right\| , \text{ or } \bar{\tau}\dot{S} = \bar{i}\,\dot{r}_1 + \bar{j}\,\dot{r}_2 + \bar{k}\,\dot{r}_3 .$$

It is known that scalar and vector productions of vectors are represented in quaternion matrices as follows (Kravets, Kravets & Kharchenko 2010):

$$\left\| \begin{matrix} \bar{a}\cdot\bar{b} \\ 0 \end{matrix} \right\| \leftrightarrow \frac{1}{2}\left(A_0 + A_0^t\right)b_0 , \quad \left\| \begin{matrix} 0 \\ \overline{a\times b} \end{matrix} \right\| \leftrightarrow \frac{1}{2}\left(A_0 - A_0^t\right)b_0 ,$$

where $A_0 = \left\| \begin{matrix} 0 & a_1 & a_2 & a_3 \\ -a_1 & 0 & -a_3 & a_2 \\ -a_2 & a_3 & 0 & -a_1 \\ -a_3 & -a_2 & a_1 & 0 \end{matrix} \right\|$,

$A_0^t = \left\| \begin{matrix} 0 & a_1 & a_2 & a_3 \\ -a_1 & 0 & a_3 & -a_2 \\ -a_2 & -a_3 & 0 & a_1 \\ -a_3 & a_2 & -a_1 & 0 \end{matrix} \right\|$, $b_0 = \left\| \begin{matrix} b_1 \\ b_2 \\ b_3 \end{matrix} \right\|$.

Then in earth reference we define:

$$\left\| \begin{matrix} \bar{V}\cdot\bar{V} \\ 0 \end{matrix} \right\| \leftrightarrow \frac{1}{2}\left(\dot{R}_0 + \dot{R}_0^t\right)\dot{r}_0 ,$$

where $\dot{R}_0 = \left\| \begin{matrix} 0 & \dot{r}_1 & \dot{r}_2 & \dot{r}_3 \\ -\dot{r}_1 & 0 & -\dot{r}_3 & \dot{r}_2 \\ -\dot{r}_2 & \dot{r}_3 & 0 & -\dot{r} \\ -\dot{r}_3 & -\dot{r}_2 & \dot{r}_1 & 0 \end{matrix} \right\|$,

$\dot{R}_0^t = \left\| \begin{matrix} 0 & \dot{r}_1 & \dot{r}_2 & \dot{r}_3 \\ -\dot{r}_1 & 0 & \dot{r}_3 & -\dot{r}_2 \\ -\dot{r}_2 & -\dot{r}_3 & 0 & \dot{r}_1 \\ -\dot{r}_3 & \dot{r}_2 & -\dot{r}_1 & 0 \end{matrix} \right\|$, $\dot{r}_0 = \left\| \begin{matrix} 0 \\ \dot{r}_1 \\ \dot{r}_2 \\ \dot{r}_3 \end{matrix} \right\|$,

or $\left\| \begin{matrix} \bar{V}\cdot\bar{V} \\ 0 \end{matrix} \right\| \leftrightarrow \left\| \begin{matrix} 0 & \dot{r}_1 & \dot{r}_2 & \dot{r}_3 \\ -\dot{r}_1 & 0 & 0 & 0 \\ -\dot{r}_2 & 0 & 0 & 0 \\ -\dot{r}_3 & 0 & 0 & 0 \end{matrix} \right\| \left\| \begin{matrix} 0 \\ \dot{r}_1 \\ \dot{r}_2 \\ \dot{r}_3 \end{matrix} \right\|$,

i.e. $\left\| \begin{matrix} \bar{V}\cdot\bar{V} \\ 0 \end{matrix} \right\| \leftrightarrow \left\| \begin{matrix} \dot{r}_1^2 + \dot{r}_2^2 + \dot{r}_3^2 \\ 0 \end{matrix} \right\|$.

Similarly in natural axes we obtain:

$$V_0 = \left\| \begin{matrix} 0 & V_\tau & V_n & V_b \\ -V_\tau & 0 & -V_b & V_n \\ -V_n & V_b & 0 & V_\tau \\ -V_b & -V_n & V_\tau & 0 \end{matrix} \right\| ,$$

$$V_0^t = \left\| \begin{matrix} 0 & V_\tau & V_n & V_b \\ -V_\tau & 0 & V_b & -V_n \\ -V_n & -V_b & 0 & V_\tau \\ -V_b & V_n & -V_\tau & 0 \end{matrix} \right\| , \quad \upsilon_0 = \left\| \begin{matrix} 0 \\ V_\tau \\ V_n \\ V_b \end{matrix} \right\| ,$$

$$\left\| \begin{matrix} \bar{V}\cdot\bar{V} \\ 0 \end{matrix} \right\| \leftrightarrow \left\| \begin{matrix} 0 & V_\tau & V_n & V_b \\ -V_\tau & 0 & 0 & 0 \\ -V_n & 0 & 0 & 0 \\ -V_b & 0 & 0 & 0 \end{matrix} \right\| \left\| \begin{matrix} 0 \\ V_\tau \\ V_n \\ V_b \end{matrix} \right\| \text{ or}$$

$$\left\| \begin{matrix} \bar{V}\cdot\bar{V} \\ 0 \end{matrix} \right\| \leftrightarrow \left\| \begin{matrix} 0 & \dot{S} & 0 & 0 \\ -\dot{S} & 0 & 0 & 0 \\ 0 & 0 & 0 & 0 \\ 0 & 0 & 0 & 0 \end{matrix} \right\| \left\| \begin{matrix} 0 \\ \dot{S} \\ 0 \\ 0 \end{matrix} \right\| \text{ i.e. } \left\| \begin{matrix} \bar{V}\cdot\bar{V} \\ 0 \end{matrix} \right\| \leftrightarrow \left\| \begin{matrix} \dot{S}^2 \\ 0 \end{matrix} \right\| .$$

Obvious equality follows:

$$\dot{S}^2 = \dot{r}_1^2 + \dot{r}_2^2 + \dot{r}_3^2 .$$

In the context of earth reference, linear acceleration vector is found out according to the specified hodograph in the form of:

$$\bar{W} = \frac{d^2\bar{r}}{dt^2} \text{ or } \bar{W} = \bar{i}\,\ddot{r}_1 + \bar{j}\,\ddot{r}_2 + \bar{k}\,\ddot{r}_3 .$$

Acceleration velocity is determined with the help of scalar production:

$$\bar{W}\cdot\bar{W} = \ddot{r}_1^2 + \ddot{r}_2^2 + \ddot{r}_3^2 .$$

Linear acceleration vector in natural axes is (Lobas 2009):

$$\bar{W} = \bar{\tau}\ddot{S} + \bar{n} K \dot{S}^2 ,$$

i.e.

$$W_\tau = \bar{\tau}\cdot\bar{W} \text{ or } W_\tau = \ddot{S} ,$$

$$W_n = \bar{n}\cdot\bar{W} \text{ or } W_n = K\dot{S}^2 ,$$

$$W_b = \bar{b}\cdot\bar{W} \text{ or } W_b = 0 ,$$

where K is curvature.
Then

$$\bar{W}\cdot\bar{W} = \ddot{S}^2 + K^2 \dot{S}^4 ,$$

29

i.e. dependence with specified hodograph takes place:

$$\ddot{S}^2 + K^2 \dot{S}^4 = \ddot{r}_1^2 + \ddot{r}_2^2 + \ddot{r}_3^2 .$$

Hodograph also determines W_r and W_n components of following closed vector form:

$$W_r = \frac{\dot{\bar{r}} \cdot \ddot{\bar{r}}}{|\dot{\bar{r}}|}, \qquad W_n = \frac{|\dot{\bar{r}} \times \ddot{\bar{r}}|}{|\dot{\bar{r}}|} .$$

Here scalar and vector products are convenient to be calculated in quaternion matrices (Kravets, Kravets & Kharchenko 2009):

$$\left\| \begin{matrix} \dot{\bar{r}} \cdot \ddot{\bar{r}} \\ 0 \end{matrix} \right\| \leftrightarrow \frac{1}{2}\left(\dot{R}_0 + \dot{R}_0^t \right) \ddot{r}_0 , \quad \left\| \begin{matrix} 0 \\ \dot{\bar{r}} \times \ddot{\bar{r}} \end{matrix} \right\| \leftrightarrow \frac{1}{2}\left(\dot{R}_0 - \dot{R}_0^t \right) \ddot{r}_0 .$$

Curvature in the system of natural trihedral coordinates is determined as:

$$K^2 = \frac{W_n^2}{\dot{S}^4} ,$$

Where $W_n^2 = \overline{W} \cdot \overline{W} - W_r^2 .$

Tangential acceleration in earthbound coordinate system is determined according to the formula:

$$W_r = \frac{1}{\dot{S}}\left(\overline{V} \cdot \overline{W} \right) \text{ or}$$

$$W_r = \frac{\dot{r}_1 \ddot{r}_1 + \dot{r}_2 \ddot{r}_2 + \dot{r}_3 \ddot{r}_3}{\left(\dot{r}_1^2 + \dot{r}_2^2 + \dot{r}_3^2 \right)^{\frac{1}{2}}} .$$

Then

$$W_n^2 = \ddot{r}_1^2 + \ddot{r}_2^2 + \ddot{r}_3^2 - \frac{\left(\dot{r}_1 \ddot{r}_1 + \dot{r}_2 \ddot{r}_2 + \dot{r}_3 \ddot{r}_3 \right)^2}{\dot{r}_1^2 + \dot{r}_2^2 + \dot{r}_3^2} .$$

Consequently in earth reference we obtain:

$$K^2 = \frac{\left(\dot{r}_1^2 + \dot{r}_2^2 + \dot{r}_3^2 \right)\left(\ddot{r}_1^2 + \ddot{r}_2^2 + \ddot{r}_3^2 \right)}{\left(\dot{r}_1^2 + \dot{r}_2^2 + \dot{r}_3^2 \right)^3} +$$

$$+ \frac{-\left(\dot{r}_1 \ddot{r}_1 + \dot{r}_2 \ddot{r}_2 + \dot{r}_3 \ddot{r}_3 \right)^2}{\left(\dot{r}_1^2 + \dot{r}_2^2 + \dot{r}_3^2 \right)^3} .$$

Equivalent vector form provides closed curvature record (Kravets, Kravets & Kharchenko 2010):

$$K^2 = \frac{\left(\dot{\bar{r}} \times \ddot{\bar{r}} \right) \cdot \left(\dot{\bar{r}} \times \ddot{\bar{r}} \right)}{\left(\dot{\bar{r}} \cdot \dot{\bar{r}} \right)^3} \text{ or}$$

$$K^2 = \frac{\dot{\bar{r}} \cdot \left[\ddot{\bar{r}} \times \left(\dot{\bar{r}} \times \ddot{\bar{r}} \right) \right]}{\left(\dot{\bar{r}} \cdot \dot{\bar{r}} \right)^3} .$$

Using determinant we also find out:

$$K^2 = \frac{\begin{vmatrix} \dot{\bar{r}} \cdot \dot{\bar{r}} & \dot{\bar{r}} \cdot \ddot{\bar{r}} \\ \ddot{\bar{r}} \cdot \dot{\bar{r}} & \ddot{\bar{r}} \cdot \ddot{\bar{r}} \end{vmatrix}}{\left(\dot{\bar{r}} \cdot \dot{\bar{r}} \right)^3} .$$

Quaternion matrices provide convenient curvature calculation (Kravets 2013):

$$K^2 = \frac{1}{4\dot{S}^6} \ddot{r}_0^t \left(\dot{R}_0 - \dot{R}_0^t \right)^2 \ddot{r}_0 , \text{ where}$$

$$\ddot{r}_0^t = \left\| 0 \ \ddot{r}_1 \ \ddot{r}_2 \ \ddot{r}_3 \right\| .$$

Here first and second products of its time components for hodograph under consideration $\bar{r}(t)$ are determined as follows:

$$\dot{r}_1 = \left\| \rho_0 \rho_1 \rho_2 \rho_3 \right\| \left[\left\| \begin{matrix} 0 \\ 1 \\ 2t \\ 3t^2 \end{matrix} \right\| \cos \omega t - \omega \left\| \begin{matrix} 1 \\ t \\ t^2 \\ t^3 \end{matrix} \right\| \sin \omega t \right] ,$$

$$\dot{r}_2 = \left\| \rho_0 \rho_1 \rho_2 \rho_3 \right\| \left[\left\| \begin{matrix} 0 \\ 1 \\ 2t \\ 3t^2 \end{matrix} \right\| \sin \omega t + \omega \left\| \begin{matrix} 1 \\ t \\ t^2 \\ t^3 \end{matrix} \right\| \cos \omega t \right] ,$$

$$\dot{r}_3 = \left\| h_0 h_1 h_2 h_3 \right\| \left\| \begin{matrix} 0 \\ 1 \\ 2t \\ 3t^2 \end{matrix} \right\|$$

$$\ddot{r}_1 = \left\| \rho_0 \rho_1 \rho_2 \rho_3 \right\| \left[\left\| \begin{matrix} -\omega^2 \\ -\omega^2 t \\ 2 - \omega^2 t^2 \\ 6t - \omega^2 t^2 \end{matrix} \right\| \cos \omega t - 2\omega \left\| \begin{matrix} 0 \\ 1 \\ 2t \\ 3t^2 \end{matrix} \right\| \sin \omega t \right] ,$$

$$\ddot{r}_2 = \left\| \rho_0 \rho_1 \rho_2 \rho_3 \right\| \left[\left\| \begin{matrix} -\omega^2 \\ -\omega^2 t \\ 2 - \omega^2 t^2 \\ 6t - \omega^2 t^2 \end{matrix} \right\| \sin \omega t + 2\omega \left\| \begin{matrix} 0 \\ 1 \\ 2t \\ 3t^2 \end{matrix} \right\| \cos \omega t \right] ,$$

$$\ddot{r}_3 = \left\| h_0 h_1 h_2 h_3 \right\| \left\| \begin{matrix} 0 \\ 0 \\ 2 \\ 6t \end{matrix} \right\| .$$

Kinetics. Mathematical model of two-wheel vehicle in terms of its spatial motion along curved lay of line is developed using non-linear differential Euler-Lagrange equations in the form of quaternion matrices (Kravets, Kravets & Kharchenko 2009). Projections of vehicle velocity vector on natural axes are assumed as quasi-velocities. Natural trihedral is taken as a bound coordinate system which pole is combined with the material point assumed as a vehicle model. Two-wheel vehicle is considered as the material point of the known mass with the applied inertial forces, gravitation force, aerodynamic forces, and the required contact moving (control) forces ensuring necessary motion mode along the specified spatial curved lay of line. Then kinetostatics equations are as follows (Kravets 2003):

$$
\frac{1}{m}\begin{Vmatrix} 0 \\ N_\tau \\ N_n \\ N_b \end{Vmatrix} = \begin{Vmatrix} 0 \\ W_\tau \\ W_n \\ 0 \end{Vmatrix} + g\,A^t\cdot {}^tA^t \begin{Vmatrix} 0 \\ 0 \\ 0 \\ 1 \end{Vmatrix} - \frac{qS}{m} R_d\cdot {}^tR_d \begin{Vmatrix} 0 \\ c_{1d} \\ c_{2d} \\ c_{3d} \end{Vmatrix}.
$$

where m is vehicle mass; g is gravity acceleration; q is velocity pressure; S is characteristic area; c_{1d}, c_{2d}, c_{3d} are aerodynamic coefficients; W_τ, W_n are quasi-accelerations; A is quaternion matrix in terms of Rodriguez-Hamilton parameters determining orientation of natural trihedral in earth reference; R_d is quaternion matrix determining orientation of aerodynamic axes relative to natural ones; and N_τ, N_n, N_b are moving forces.

Kinematic correlations in quaternion matrices closing the given kinetostatics equations are as follows (Kravets, Kravets & Kharchenko 2009):

$$
\begin{Vmatrix} 0 \\ \dot r_1 \\ \dot r_2 \\ \dot r_3 \end{Vmatrix} = A\cdot {}^tA \begin{Vmatrix} 0 \\ V_\tau \\ 0 \\ 0 \end{Vmatrix}, \quad \begin{Vmatrix} 0 \\ V_\tau \\ 0 \\ 0 \end{Vmatrix} = A^t\cdot {}^tA^t \begin{Vmatrix} 0 \\ \dot r_1 \\ \dot r_2 \\ \dot r_3 \end{Vmatrix}.
$$

Statics. Obtained resulting moving force $\left(\overline{N}\right)$ which provides motion of two-wheel vehicle along specified lay of line in the known mode is represented as a system of two equivalent contact control forces $\left(\overline{F}_1, \overline{F}_2\right)$ to be determined. Following figure shows them:

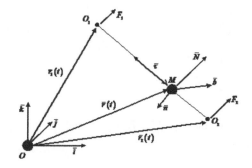

Figure 2.

Here reference points are given in movable natural axes: $O_1M = l_1$, $O_2M = l_2$.

Then according to Varignon theorem (Lobas 2009) we obtain:

$$\overline{r}_1\times\overline{F}_1 + \overline{r}_2\times\overline{F}_2 = \overline{r}\times\overline{N},$$

where $\overline{r}_1 = \overline{r} + \overline{\tau}l_1$, $\overline{r}_2 = \overline{r} - \overline{\tau}l_2$.

In particular, if $\overline{r}_2 = 0$, then $\overline{r} = \overline{\tau}l_2$, $\overline{r}_1 = \left(l_1 + l_2\right)\overline{\tau}$ and $\left(l_1 + l_2\right)\overline{\tau}\times\overline{F}_1 = l_2\overline{\tau}\times\overline{N}$.

Hence:

$$F_{1n} = \frac{l_2}{l_1 + l_2}N_n, \qquad F_{1b} = \frac{l_2}{l_1 + l_2}N_b,$$

and

$$\frac{F_{1n}}{N_n} = \frac{F_{1b}}{N_b} = \frac{l_2}{l_1 + l_2}$$

or parallelism condition:

$$\frac{F_{1n}}{F_{1b}} = \frac{N_n}{N_b}.$$

If $\overline{r}_1 = 0$, then $\overline{r} = -l_1\overline{\tau}$, $\overline{r}_2 = -\left(l_1 + l_2\right)\overline{\tau}$ and $\left(l_1 + l_2\right)\overline{\tau}\times\overline{F}_2 = l_1\overline{\tau}\times\overline{N}$.

Hence:

$$F_{2n} = \frac{l_1}{l_1 + l_2}N_n, \qquad F_{2b} = \frac{l_1}{l_1 + l_2}N_b,$$

and

$$\frac{F_{2n}}{N_n} = \frac{F_{2b}}{N_b} = \frac{l_1}{l_1 + l_2}$$

or parallelism condition:

31

$$\frac{F_{2n}}{F_{2b}} = \frac{N_n}{N_b} .$$

Consequently:

$$\frac{F_{1n}}{F_{1b}} = \frac{F_{2n}}{F_{2b}} = \frac{N_n}{N_b} .$$

If $\bar{r} = 0$, then $\bar{r_1} = l_1 \bar{\tau}$, $\bar{r_2} = -l_2 \bar{\tau}$ and $l_1 \bar{\tau} \times \bar{F_1} - l_2 \bar{\tau} \times \bar{F_2} = 0$.

It follows:

$$\frac{F_{1b}}{F_{2b}} = \frac{l_2}{l_1}, \frac{F_{1n}}{F_{2n}} = \frac{l_2}{l_1}$$

or

$$\frac{F_{1n}}{F_{2n}} = \frac{F_{1b}}{F_{2b}} = \frac{l_2}{l_1},$$

i.e. pre-determined parallelism condition

$$\frac{F_{1n}}{F_{1b}} = \frac{F_{2n}}{F_{2b}} .$$

Static invariants are required to verify obtained results. In particular, within normal plane of natural trihedral static invariant one:

$$F_{1n} + F_{2n} = N_n , \qquad F_{1b} + F_{2b} = N_b$$

is satisfied equally. Condition $F_{1\tau} + F_{2\tau} = N_\tau$ is used to specify required torque of traction wheel at definite resistance of driven one and required mode of vehicle motion along the specified lay of line. Static invariant two:

$$\bar{F_{1n}} \cdot \left(\bar{N} \times \bar{r_1} \right) + \bar{F_2} \cdot \left(\bar{N} \times \bar{r_2} \right) = 0$$

results in parallelism condition

$$\frac{F_{1n}}{F_{1b}} = \frac{F_{2n}}{F_{2b}} = \frac{N_n}{N_b} , \quad \text{and} \quad \frac{l_1 F_{1n} - l_2 F_{2n}}{l_1 F_{1b} - l_2 F_{2b}} = \frac{N_n}{N_b} .$$

Thus, analytical solution determining equivalent contact of moving (control) forces for tandem-type two-wheel vehicle at various motion modes along spatial curved lay of line under the effect of gravity and aerodynamic forces is obtained. The closed vector dependences are shown in the form of quaternion matrices providing efficient computational algorithms.

REFERENCES

Khachaturov A.A. 1976. *Dynamics of a road-rail-car-driver system* (in Russian). Moscow: Mechanical Engineering: 535.

Kravets T.V. 2003. *Definition of control forces and moments when driving asymmetric aircraft to program trajectory of complex spatial configuration* (in Russian). Technical Mechanics, Issue 1: 60-65.

Kravets V.V., Kravets T.V. & Kharchenko A.V. 2009. *Using quaternion matrices to describe the kinematics and nonlinear dynamics of an asymmetric rigid body.* Int. Applied Mechanics, Issue 45, Vol.2: 223-232.

Kravets V.V., Kravets T.V. & Kharchenko A.V. 2010. *The method of the matrix representation of vector algebra multiplicative compositions* (in Russian). East European Journal of advanced technologies, Issue 3/6 (45): 12-16.

Kravets T.V. 2013. *On the use of quaternion matrices in the analytical and computational solid mechanics* (in Russian). Technical Mechanics, Issue 3: 91-102.

Lobas L.G. 2009. *Theoretical Mechanics* (in Ukrainian). Kyiv: DETYT: 407.

Martunjuk A.A., Lobas L.G. & Nikitina N.V. 1981. *Dynamics and stability of motion wheelsets transport vehicles* (in Russian). Kyiv: Technics: 223.

Independent power supply of menage objects based on biosolid oxide fuel systems

O. Beshta
State Higher Educational Institution "National Mining University", Dnipropetrovs'k, Ukraine

V. Fedoreyko, A. Palchyk & N. Burega
V. Hnatiuk Ternopil National Pedagogical University, Ternopil, Ukraine

ABSTRACT: Feasibility justification of microbiotechnology implementing in an autonomous power supply system based on solid oxide fuel cells for laminar renewable energy inflow. Investigation of guaranteed power supply is based on the laws of conservation of mass, electrical engineering, electrochemistry; on the biological processes of photosynthesis and anaerobic fermentation using data obtained from simulation methods and physical modeling. The current status of autonomous power supply, industrial photobioreactors, anaerobic fermenters and fuel cell systems is considered; analysis of simulation model operation of autonomous electrotechnological energy supply based on fuel cells and alternative energy sources is made; the technological flow chart of the solid oxide fuel cell with photoanaerobic bioreactor is developed. Feasibility of using biological technology of anaerobic fermentation in power complex of autonomous energy supply systems is proved. Structural and technological schemes of autonomous power supply complex of manage objects based on fuel cells and photoanaerobic bioreactors are developed.

1 INTRODUCTION

Energy consumption is growing worldwide and resources of the biosphere are permanently reducing resulting in significant negative anthropogenic impact. To a large extent ecological problems take place because of extensive emission and intrusion of technogenesis into material and energy balance of the planet by release of great amount of carbon dioxide. Carbon dioxide emission are connected with necessity to generate electric, heat or mechanical energy in various power systems.

Alternative types of fuel are the main catalysts of new global tendencies in energy sector market due to reduction of mineral reserves, high dependence on import of oil and natural gas and structure change of agro-industrial production. Specialists all over the world direct their considerable scientific and industrial potential to reduction of energy dependency from fossil fuel. Full-scale using of energy of biomass, wind, microorganisms, water and solar energy will enable to stabilize material and energy balance of the planet that comes up now to a risky level.

In our days technologies are widely developing that are directed to decreasing greenhouse gases emission and connected with developing of autonomous power supply complexes based on renewable energy sources. But using of energy of wind and solar radiation is quite complicated due to its stochastic nature and irregular consumption. So in autonomous power supply complexes there is a need to accumulate excesses of energy for the period when it will become inaccessible or its inflow will be not sufficient to provide required parameters of power supply. For the operation during a year this approach stipulates the using of large chemical energy storages and electrochemical storage batteries of high capacity.

One of the promising trends developing in our days is using of microorganisms in power complexes for generation of biogas, biodiesel and hydrogen.

2 PURPOSE

Feasibility justification of microbiotechnology implementing in an autonomous power supply system based on solid oxide fuel cells for laminar renewable energy inflow.

3 MATERIALS AND METHODS OF RESEARCH

The research of guaranteed energy supply is

based on the laws of conservation of mass, electrical engineering, electrochemistry; on biological processes of photosynthesis and anaerobic fermentation using data obtained from simulation methods and physical modeling.

4 RESULTS OF RESEARCH

Growing microalgae enables to solve one of the main human problems – decreasing of carbon dioxide in atmosphere. 1 ton of phytoplankton during its growth absorbs 1,6 tons of CO_2 and generates 1,2 tons of oxigen. By the way unicellular organisms are able to produce and amass triglycerides. A constituent part of these oils in one cell of microalga can be from 35 to 70% and sometimes above 80% (Botryococcus braunii) (Thompson, Advisor 2010) of cell total mass. These oils are the basic material for producing the «third generation» biofuel because they make it possible to get up to 94000 kg of oil per one hectare.

Cultivation of microalgae is carried out by mankind during centuries mainly for food. The using of dry biomass of chlorella and spirulina is widely extended in livestock business and fishery. Its typical and very important genetic feature is rapid reproduction. The frequency of mass duplication is individual for every species: the slowest – 1-2 times per a day and the quickest – 8-10 times.

In 2010 Algenol Biofuels and Dow Chemical started the construction of 24-acres experimental plant in Texas that will consume 1,8 tons of industrial enterprises CO_2 during a day and generate 100 000 gallons of ethanol per year. In case of success it is planned to construct industrial «treatment» plant near Puerto Libertad power plant in Sonora desert (Mexico). Scale of production will be 170 000 acres, it will absorb 6 000 000 tons of carbon dioxide generating 1 billion gallons of ethanol per year. As for 2007 the biogas production in EU countries was 5 900 kg of oil equivalent and as European Biomass Association (AEBIOM) estimates it will be 40 000 kg up to 2020 that is 10% of gas consumption by EU countries (Thompson, Advisor 2010). In early 2006, several companies have announced the start of construction of plants to produce biodiesel from algae: Global Green Solutions (Canada); Corporation (USA) with capacity of 900 million gallons of biodiesel per year.

On the other hand, one of the ways to reduce emissions of carbon dioxide in the atmosphere may be improving energy efficiency of electrical energy generation. A number of new technologies are being developed in this area, one of which is the use of solid oxide (SOFC - Solid Oxid Fuell Cell) or proton exchange (PEMFC - Proton exchange membrane fuel cells) fuel cells. Their peculiarity is that they can during one technological stage convert chemical fuel (H_2 and CH_4) into electricity with extremely high efficiency, which is about 70%.

The use of fuel cells with biotechnology can allow the construction of a self-sufficient autonomous power supply system of menage objects with zero greenhouse gas emissions and the possibility of disposing of biological waste. This approach is implemented by the anaerobic fermentation of biomass, generation of biogas from it, chemical energy of which is used to produce electricity and heat. As for combustion products (CO_2) they are re-used to generate biomass in photoreactor (Fig. 1). Therefore, this system can provide customers a wide range of energy (heat and electrical) and energy carriers (biogas and biodiesel).

Application of fuel cells is caused by high efficiency (the ability to use hydrogen as a fuel), as its principle of operation is the oxidation of hydrogen or methane by ionic conductivity of polymeric and ceramic membranes. When using proton exchange fuel cell, methane is converted into hydrogen which is then filtered and fed into the stack.

Fuel cell converts chemical fuel (H_2 and CH_4) in one technological stage into electricity with extremely high efficiency, which is about 70% – much higher than internal combustion engines and steam turbine (20% and 45%).

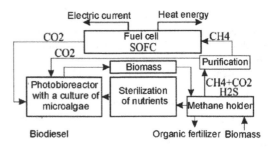

Figure 1. Structural diagram of a power system based on the fuel cell and photobioreactor

Fuel cell voltage depends on Nernst polarization potential of the cathode, anode and electrolyte internal resistance (Strasser 1990).

$$V = E_0 - iR_w - \eta_{cathode} - \eta_{anode} \qquad (1)$$

Nernst equation depends on the type of chemical reaction. For hydrogen equation is:

$$E = E^\circ + (\frac{RT}{2F})\ln[\frac{P_{H_2}}{P_{H_2O}}] + (\frac{RT}{2F})\ln[P^{\frac{1}{2}}_{O_2}] \qquad (2)$$

When using methane as fuel, the equation becomes:

$$CH_4 + 2O_2 \rightarrow 2H_2O + CO_2 \tag{3}$$

$$E = E^o + (\frac{RT}{8F})\ln[\frac{P_{CH_4}}{P^2_{H_2O}P_{CO_2}}] + (\frac{RT}{8F})\ln[P^2_{O_2}] \tag{4}$$

The ability to use hydrogen as an energy carrier allows you to receive and store the excess of renewable energy during the annual cycle. This approach provides the menage object with energy supply from three independent sources (Fig. 2): wind flow, biomass and solar radiation.

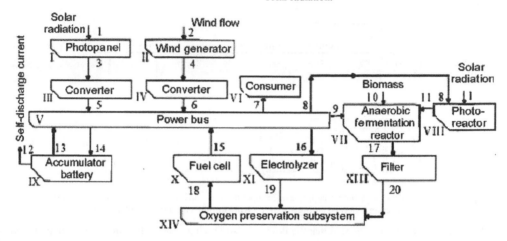

Figure 2. A model of autonomous power supply based on fuel cells and renewable energy sources

Using Matlab Simulink, developed mathematical model and conducted experimental studies it was designed and researched the simulation model of autonomous power supply of a menage object and the 1st kind habitation on the basis of the fuel cell, wind turbine and photovoltaic cells (Fig. 3). Imitation modeling showed necessity to create a powerful system of hydrogen storage (up to 300m3) (Fig. 4 a) for the annual period (Fig. 4, B), which is associated with stochastic inflow of energy from renewable sources (Fig. 4, c and d).

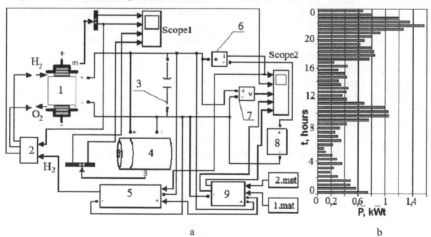

a   b

Figure 3. a) Simulation model of autonomous power supply system: 1 - fuel cell, 2 - control unit for fuel supply, 3 - lead-acid battery, 4 - electrolyzer, 5 - amperemeter, 6 - voltmeter, 7 – block of operation load imitation of the first kind housing, 8 – imitation block of photovoltaic power battery and wind generator; b) - the daily load distribution of a menage object.

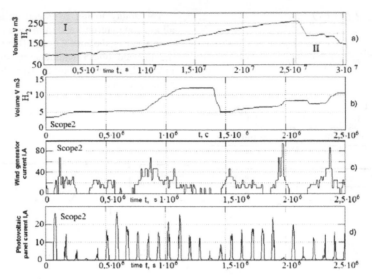

Figure 4. Parameters of autonomous power supply system operation: a - the amount of hydrogen in a storage during a year; b - the amount of hydrogen in a storage during a year; c - wind turbine current; d - photopanel current.

Based on the simulation it can be said that the flow of energy throughout the year from renewable sources is uneven and requires more laminar renewable source.

The use of energy of biomass for autonomous power supply system is caused by laminar nature of its arrival, and thus makes it possible to reduce the power of the wind generator, photopanel and size of energy carrier storage sites. Therefore to create mathematical and imitation models the analysis of systems of biological reactors for photosynthesis and anaerobic fermentation is required.

Over the past few decades, a large part of research in the area of cultivation was aimed at developing open large-scale industrial facilities (ponds), that are currently exploited worldwide. However for use in the energy sector, these systems can not be applied due to bacterial culture contamination, climate and seasonal dependence. Therefore, to eliminate these deficiencies the closed systems of growing algae – photo bioreactors are used. These systems provide solar and artificial light, can control all aspects of the life cycle of algae: temperature, pH, redox potential of the medium, the concentration of dissolved oxygen, level of CO_2 and nutrients, velocity of the fluid, the intensity of mixing.

Photo bioreactors operate in accumulation mode with feeding or continuous cultivation – to support a given density of the suspension by periodic removal of algae and adding nutrient solution. There are three types of structures: tubular reactors (horizontal, vertical, and spiral), plane-parallel ditch, photobioreactors on the basis of coaxial cylinders.

Analyzing the structure of existing photo bioreactors we have identified a plane-parallel ditch and tubular system. It is they in our opinion are the most promising for research and application in the industrial version.

Plane-parallel ditch has several advantages: ease of manufacturing, large amount of cultivated environment that solves the problem of even natural and artificial lighting which saturates photosynthesis. Despite its simplicity, this type of reactor is particularly useful in the study of the physiological characteristics of unicellular microalgae, cyanobacteria and allows to identify the main energy characteristics of growth. Since microalgae have a high content of lipids (fats), the main bulk of them will be at the top. Gathering of algae is carried out through the hole for harvest exit. This is possible by the input of water with nutrient medium into a photobioreactor that raises the level of suspension and removes microorganisms from cultivation environment.

The main drawback of such a photo bioreactor is a process of gas mass transfer, which in this type has low limits (high feed rate forms large cavities and gas leaves with little or no mixing with a suspension). This makes impossible the process of increasing the rate of flow of the gas mixture which in turn complicates the process of continuous cultivation.

In developing plane photo bioreactor it should be considered its locations and position: vertical or horizontal. During the study the plane-parallel ditch was placed horizontally changing the position of axis from north to south and vertically - "East-

36

West". Obtained results demonstrated that for the horizontal photobioreactors intercepted radiation ranged from 11 to 30 MJ/m^2 during a day, while for vertical "east-west" orientation the total absorbed radiation was between 13 to 29 MJ/m^2 during a day. The data showed that in summertime the amount of solar radiation received by two types are almost the same. However in winter vertical photo bioreactor with "east-west" orientation intercepted 17% more radiation than a horizontal surface, but in summer – 3% less (Sierra, Acien, Fernandez, Garcıa, Gonzalez 2008).

A common feature of tubular systems is presence of two modules: a light and gas mass transfer modules. Photosynthesis mainly takes place in the system of transparent tubes that are connected to the gas mass transfer reservoir. The suspension of algae constantly circulates through transparent tubes from the reservoir to the gas mass transfer exchanger. Fresh nutrient medium and gas mixture come in the gas mass-transfer exchanger and from there with the

use of mixers or circulation pumps – in the tubular part of the reactor. The selection of algae portion from tubular photo bioreactor is conducted similar to plane-parallel ditch: increasing the level of the liquid and selection through the top nozzle that provides operation in continuous mode.

The system design should perform the functions of measurement and control of temperature (support by the way of heat exchange between the nutrient medium and injected heat carrier), light (artificial and external), the concentration of acidity (pH level, which is regulated by adding the appropriate amount of CO_2 or by appropriate salt solution) (Fig. 5).

At cultivation of algae the photobioreactor should be responsible for the biological purity of culture, for keeping away bacteria and other microorganisms and if necessary – to suppress them. Usually ultraviolet lighting of medium wavelength, antibiotics or bacteriophages are used for extermination of bacteria.

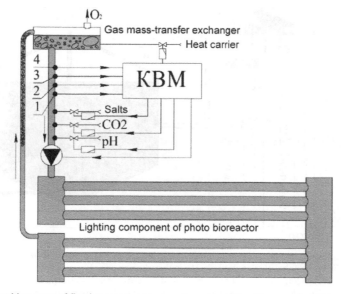

Figure 5. Tubular photobioreactor of flowing type 1 - a temperature sensor; 2 - pH sensor; 3 - light sensor; 4 - sensor of microalgae concentration.

Using transparent polyvinylchloride and fluoroplastic tubes it is possible to provide different forms for photobioreactors. Due to small diameter of the tubes, the culture of microalgae can be well illuminated even in the center that maximizes access for photosynthesis. Since photobioreactor is a symbiosis of technological and biological object its successful implementation depends entirely on engineering solutions in environment intelligent control.

For laboratory studies we developed a cylindrical pilot project of photobioreactor (Palchik, Burega 2013) that has the volume of 3 liters. Its peculiarity is the presence of a parabolic reflector that focuses solar radiation and additional artificial lighting. Active growth of microalgae creates pillar in the light zone thus limiting the penetration of solar radiation inside the photo bioreactor that restricts in manufacture the pipe diameter and respectively reduces its volume. Placing backlight inside the reactor with three types of LEDs has improved illumination of dark areas and decreased formation of a film on the walls that reduces the natural light transmission coefficient. Microorganisms move to the lighter parts of photobioreactor: during a day – to the outer walls and at night – to the center of the bulb (where the artificial light is).

After analyzing the effects of different intensity and color of artificial light we used three types of LEDs (Light Emitting Diode): Blue – increases the number of lipids in the composition of Chlorella, red – promotes growth and reduces biomass fermentation the nutrient medium and white.

On the basis of the developed photo bioreactor the work is being done on its systems automation and methods are being examined of reactions assessment of the biological system (Chlorella and Anabena) to changes in feeding environment, temperature and lighting regimes.

Designed photo bioreactor is intended for production of microalgae as carbon dioxide filters in industrial production and also to generate biomass which is used as a food protein, raw materials for cosmetics and medicine, bio-diesel.

A fundamentally new method in our system is the use of anaerobic fermentation by systems, fuel cells and microalgae that will allow to build a new power system of zero-kind housing. After analyzing biogas facilities which are now used commercially, we propose to use БГУ – 05 for subsequent studies (Burga, Riger, Vayland, Schreder 2010). This installation has several advantages for construction: reasonable price, reasonable size of the reactor and gas storage, a large area of organic fertilizers application.

Based on this installation the technological power system was developed (Fig. 6) which has a closed circle of carbon dioxide circulation (in the photoanaerobic bioreactor). It is assumed the use of household food waste and livestock and crop residues for biogas generation which is purified from admixtures of hydrogen sulfide and carbon dioxide.

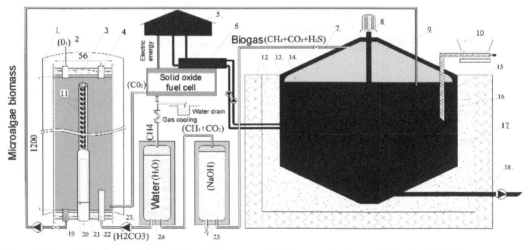

Figure 6. The technological system of using of photoanaerobic bioreactor based on solid oxide fuel cell: 1 – parabolic reflector, 2 – 4.5 mm thickness polycarbonate tube, 3, 21 – polycarbonate cap, 4 – an opening for food medium input, 5 - menage objects, 6 - heat exchanger, 7 - removal of biogas out of the fermenter, 8 - electric motor, 9 - microalgae biomass drain out of photoreactor, 10 – charging biomass auger, 11 - LED Strips, 12 - heat insulation (concrete blocks and sand), 13 – a body of methane holder, 14 - storage of biogas, 15 - mixer, 16 - heat exchange tubes 17 - organic biomass, 18 – an opening for organic residues unloading, 19 - an opening for biomass drain, 20 – a heater with temperature controller, 22 – an opening for the water supply with carbon dioxide, 23 – an opening for CO_2 supply, 24 - a reservoir for carbon dioxide purification, 25 - a reservoir for hydrogen sulfide purification.

The resulting biogas is used to generate electricity and heat. Biomass of microorganisms serves as a biological material.

The proposed system will allow to minimize carbon dioxide emissions into the atmosphere and smooth over the uneven flow of renewable energy for autonomous menage objects.

5 CONCLUSIONS

1. The use of solid oxide fuel cells is a promising way to improve the efficiency of electricity generation obtained with biogas (by anaerobic fermentation) from 20-45% up to 70%. The use of photobioreactors provides filtration of biogas from CO_2 (20%-50%) and allows to obtain additional lipids mass up to 94 000 kg per hectare as raw

material for further fermentation or synthesis of biodiesel.

2. Based on the imitation modeling of electrotechnological complex of autonomous power supply based on renewable sources it was validated the feasibility of implementing microbiotechnology in an autonomous power supply system on the basis of solid oxide fuel cells.

3. The technological scheme of power supply of menage object is proposed through the use of photoanaerobic bioreactor and solid oxide fuel cells.

REFERENCES

Thompson R.W., Advisor M. 2010. *Algae Biodiesel.* An Interactive Qualifying Project Report submitted to the Faculty of Worcester polytechnic institute: 47.

Strasser K. 1990. *An investigation on the performance optimization of an alkaline fuel cell.* Journal of Power Sources: 152–153.

Sierra E., Acien F.G., Fernandez J.M., Garcia J.L., Gonzalez C. 2008. *Characterization of a flat plate photobioreactor for the production of microalgae.* Almeria: Chemical Engineering Journal, Issue 138: 136-147.

Palchik A.O., Burega N.V. 2013. *The using of photobioreactor in alternative power supply systems* (in Russian). Works of Taurian State Agrotechnological University, Issue 13, Vol.5: 41–47.

Burga G., Riger K., Vayland P., Schreder J. 2010. *Biogas on the basis of renewable raw material. Comparative analysis of sixty one installations for biogas production in Germany.* Publication of renewable resources special agency, Issue 1: 115.

The use of asymmetric power supply during the procedure of equivalent circuit parameters identification of squirrel-cage induction motor

O. Beshta & A. Semin
State Higher Educational Institution "National Mining University", Dnipropetrovs'k, Ukraine

ABSTRACT: Objective is to get the substantiation of carrying out the identification procedure of induction motor equivalent circuit parameters when the rotor is stationary. Methods for describing of electromagnetic processes known from the theory of electrical machines and method for solving of nonlinear equations systems are used. Expressions are obtained that allow to determine the parameters of the induction motor equivalent circuit and use as an input the experimental data of AC and DC asymmetric supply of stator. A method is proposed which allows to identify the parameters of the equivalent circuit of squirrel cage induction motor with stationary rotor. For this purpose DC and AC power supply are used. Proposed expressions can serve as a basis for the development of algorithms and software for microprocessor systems of AC drives with induction motors for the identification of equivalent circuit parameters

1 INTRODUCTION

To ensure the required quality of transients of drive with squirrel-cage induction motor it is necessary to know the parameters of the motor equivalent circuit.

Many papers were published about parameters identification. In general authors solve two different problems. The first one concerns identification of parameters for initial adjusting of the control system (self-commissioning stage). The second problem is connected with parameters identification for continuous self-tuning during the drive operation. In this paper we deal with the first problem.

Traditional test methods (no-load operation test, short-circuit test) are not practicable, because often there is no enough test time or no load machine is available. So it is a strong necessity to use other methods for parameters identification.

The identification procedure is performed for the first time of drive running. For automatic identification the induction machine should be fed by frequency converter directly, when the electric motor is subjected to test actions based on which parameters are defined.

As for parameter estimation at standstill we can find in the literature three approaches: sinusoidal voltage excitation (different frequencies may be used) of the machine (Rodkin & Zdor 1998; Rodkin & Romashikhin 2009; Kalinov & Cherniy 2004), voltage step excitation (Peixoto & Seixas 1999) and

parameters estimation from measured frequency response of the machine (Conte, Pereira, Haffner, Scharlau, Campestrini & Fehlberg 2003). Each of them has its advantages and disadvantages concerning precision, time and recourses required for calculation or required equipment.

Machine tool may cause inability to obtain no-load operation mode which is used for parameters identification. In such situations it is convenient to use asymmetrical power supply of induction motor. Thus, by applying voltage to two stator phases instead of three ones the rotor will remain stationary (standstill) and the electromagnetic torque will be equal to zero.

The aim of the paper is to get the possibility of justification of carrying out the identification procedure of induction motor equivalent circuit parameters when the rotor is stationary. What is more, the method should have low sensitivity to errors in the initial experimental data , require little time and computational resources to make all necessary calculations.

2 THE USE OF DC STATOR SUPPLY

Consider the case of DC stator supply. Let phases A and B are supplied by DC voltage of value U. It can be assumed that phase A is supplied with the voltage source U/2, and Phase B - with a voltage source –U/2. This will satisfy the condition ua + ub

+ uc = 0. The space vector will be stationary in the complex plane and equal to:

$$\tilde{U} = \frac{2}{3}\left(\frac{U}{2} + \left(-\frac{U}{2}\right)e^{j120^0}\right) = \frac{U}{\sqrt{3}}e^{-j30^0}.$$

The space vector equations of the stator and rotor circuits and voltages in the fixed reference frame are:

$$\begin{cases} \tilde{U}_S = \tilde{I}_S \cdot R_1 + L_1 \dfrac{d\tilde{I}_S}{dt} + L_{12}\dfrac{d\tilde{I}_R}{dt}, \\ 0 = \tilde{I}_R \cdot R_2 + L_{12}\dfrac{d\tilde{I}_S}{dt} + L_2 \dfrac{d\tilde{I}_R}{dt}, \end{cases} \quad (1)$$

where:

R_1, R_2, –resistances of stator and rotor phase windings;

L_1, L_2, L_{12} –stator and rotor phase inductances and magnetizing branch inductance.

Projecting the equation on the real axis of the complex plane α, which coincides with the axis of the stator phase A, we get (α index in the notation of currents and voltage is omitted):

$$\begin{cases} u_S = i_S \cdot R_1 + L_1 \dfrac{di_S}{dt} + L_{12}\dfrac{di_R}{dt}, \\ 0 = i_R \cdot R_2 + L_{12}\dfrac{di_S}{dt} + L_2 \dfrac{di_R}{dt}, \end{cases} \quad (2)$$

where: u_s and i_s correspond to voltage and current of stator phase A. From the system (2) we obtain the following equation relating u_s and i_s to motor parameters:

$$\left(\frac{L_1 L_2 - L_{12}}{R_2}\right)\frac{d^2 i_S}{dt^2} + \left(\frac{L_1 R_2 + R_1 L_2}{R_2}\right)\frac{di_S}{dt} +$$

$$+ R_1 i_S = u_S + \frac{du_S}{dt}\frac{L_2}{R_2}. \quad (3)$$

This is a non-homogeneous second order differential equation with constant coefficients. Its solution consists of general solution of the corresponding homogeneous equation and a particular solution of the inhomogeneous equation. Solving (3) with the zero initial conditions, and $u_s = const$, we get:

$$i_S = \frac{u_S}{R_1} + C1 \cdot e^{-t/T1} + C2 \cdot e^{-t/T2}, \quad (4)$$

where

$$T1 = \frac{2L_{12}(L1\sigma + L2\sigma)}{L1\sigma \cdot R2 + L2\sigma \cdot R1},$$

$$T2 = \frac{L1\sigma + L2\sigma}{R1 + R2},$$

$$C1 = \frac{u_S}{R1} \cdot \frac{T1}{(T2 - T1)},$$

$$C2 = -\frac{u_S}{R1} \cdot \frac{T2}{(T2 - T1)},$$

$\dfrac{u_S}{R_1}$ – is the current i_s value at the end of the transient process,

$$u_S = const = \text{Re}(\tilde{U}_S) = \frac{U}{2}.$$

Experimental data of the current curve will have noise. To reduce its effect during the identification procedure it is convenient to find area of a figure between a current value equal to $\dfrac{u_S}{R_1}$ and current curve i_s. This area will be:

$$S = \int_0^\infty \left(\frac{u_S}{R1} - i_S\right)dt = \frac{u_S}{R1}(T1 + T2). \quad (5)$$

The ratio $T1/T2$ (and therefore $C1/|C2|$) for induction motors of 0.01-100kW is about 25-120 [6]. Thus, the transient process will be determined mainly by component $C1 \cdot e^{-t/T1}$.

3 THE USE OF AC STATOR SUPPLY

Let us consider asymmetrical AC power supply of stator phases A and B. We assume that the transition processes have been completed and we have a steady state. It is well known that asymmetrical system of voltages and currents can be represented as the sum of positive, negative and zero sequence components (zero sequence of currents is absent if the stator windings are connected in star).

We consider that phase A has a voltage source $\dfrac{Um}{2} \cdot \sin(\omega t)$, and phase B - voltage source

$\dfrac{-Um}{2} \cdot \sin(\omega t)$. They correspond to complex values:

$$\underline{U}_A = -\underline{U}_B = \frac{Um}{2} e^{j0^0}, \ \underline{U}_C = 0.$$

The complex values of stator phase currents are:

$$\underline{I}_A = -\underline{I}_B = \underline{I}_1, \ \underline{I}_C = 0.$$

Positive phase-sequence components of the stator voltage and current according to (Bessonov 1996) are:

$$\underline{U}_{11} = \frac{Um}{6}\left(1 - e^{j120^0}\right), \tag{6}$$

$$\underline{I}_{11} = \frac{I_1}{3}\left(1 - e^{j120^0}\right). \tag{7}$$

Expressions (6) and (7) allow to obtain the voltage and current in the symmetric mode for slip s = 1, using the experimental data of asymmetrical stator power supply.

Having determined the ratio of voltage and current, calculated from (6) and (7) we obtain the value of the input resistance and reactance of the equivalent circuit of the motor (Beshta & Semin 2012).

$$\underline{Z} = Rvx + jXvx,$$

where

$$Rvx = R_1 + \frac{(\omega \cdot Lm)^2 \cdot R_2}{R_2^2 + (\omega \cdot Lm + \omega \cdot L_2\sigma)^2},$$

$$Xvx = \omega \cdot L_1\sigma +$$

$$+ \frac{\omega \cdot Lm \cdot \omega \cdot L_2\sigma \cdot (\omega \cdot Lm + \omega \cdot L_2\sigma) + \omega \cdot Lm \cdot R_2^2}{R_2^2 + (\omega \cdot Lm + \omega \cdot L_2\sigma)^2}$$

As it is shown in (Kalinov & Cherniy 2004) the above expressions have very important feature. They can be rewritten as:

$$Rvx = R_1 + \frac{\omega^2 \cdot C \cdot (B - A)^2}{C^2 + \omega^2 \cdot (B - A)^2},$$

$$Xvx = \omega \cdot B - \omega^3 \cdot \frac{(B - A)^3}{C^2 + \omega^2 \cdot (B - A)^2},$$

where

$$L_1\sigma + \frac{Lm \cdot L_2\sigma}{Lm + L_2\sigma} = A,$$

$$L_1\sigma + Lm = B,$$

$$\frac{Lm^2 \cdot R_2}{(Lm + L_2\sigma)^2} = C.$$

This indicates that supplying stator with different frequencies voltage we can obtain four parameters maximum.

If we take equal leakage inductances of the stator and rotor ($L_1\sigma = L_2\sigma = L\sigma$), then the equations system for the identification of equivalent circuit parameters of induction motor becomes:

$$\begin{cases} \dfrac{4L_{12}}{R_1 + R_2} + \dfrac{2L\sigma}{R_1 + R_2} = K1, \\[2mm] R_1 + \dfrac{(\omega \cdot L_{12})^2 \cdot R_2}{R_2^2 + (\omega \cdot L_{12} + \omega \cdot L\sigma)^2} = K2, \\[2mm] \omega \cdot L\sigma + \dfrac{\omega \cdot L_{12} \cdot \omega \cdot L\sigma \cdot (\omega \cdot L_{12} + \omega \cdot L\sigma) + \omega \cdot L_{12} \cdot R_2^2}{R_2^2 + (\omega \cdot L_{12} + \omega \cdot L\sigma)^2} = \\[2mm] = K3. \end{cases} \tag{8}$$

The first equation of system (8) is obtained from (5), taking into account $L_1\sigma = L_2\sigma = L\sigma$. The constants K2,K3 are determined from the ratio of voltage and current, calculated from (6) and (7) (input resistance and reactance of the equivalent circuit of the motor (Kalinov & Cherniy 2004)) on the basis of experimental data. The stator resistance is measured at DC mode.

Systems (8) can be solved with one of the numerical method used for systems of nonlinear equations (for example, Newton's method).

4 INPUT DATA ERROR ESTIMATION

It should be noted that experimentally obtained values of the currents and voltages used for identification comprise error. The effect of this error in the articles on this subject is not discussed in detail.

In industry there are three widely used methods for measuring the current: resistive, Hall-effect and current transformers. Each method has its advantages and disadvantages which determine the area of their application. Error for various methods of current measuring of modern sensors, according to (Danilov 2004), are: resistive -< 1% Hall-effect - < 10%, current transformers - < 5%.

Evaluation of effect of errors of in the input data can be performed by the following algorithm:

1) using pre-known parameters of the equivalent circuit calculate the current and voltage of the stator;

2) change the obtained values of current and voltage according to the value of error;

3) using the currents and voltages comprising the error as the initial data for finding the parameters of the equivalent circuit;

4) determination of the ratio of the obtained values of the parameters to the original, used for calculations in step 1).

To disseminate the results to the motors of different power, equations have been considered in relative units for 3 sets of equivalent circuit parameters corresponding to motors 0.01 kW, 1 kW and 100 kW. Parameters values are shown in Table 1 (Ivanov-Smolenskiy 1980).

Table 1. Equivalent circuit parameters (in relative units)

Parameters values	Power		
	0,01 kW	1 kW	100 kW
Xm	1,25	2,5	4
Rm	0,4	0,5	0,6
R_1	0,1	0,05	0,02
R_2	0,12	0,06	0,03
X_1	0,1	0,09	0,07
X_2	0,15	0,12	0,085

The value of possible error in the input data is assumed to be ± 1%. This value was used for correction of values of K1-K3 in (8). Solutions of the system of nonlinear equations (8) for different combinations of errors were found using computational tools of MathCad. The initial values of the unknown parameters of the equivalent circuit (at which to start an iterative search) were taken quite close to the original values presented in Table 1. Staror voltage frequency accept equal to 5 Hz. In all cases, it was considered that stator resistance R1 was already known.

Initial experimental data for determination of the motor parameters is recommended to obtain:

a) by applying a DC voltage to two phases of the stator (for use in the first equation (8));

b) by applying an alternating voltage to two phases of the stator (for use in the second and third equations of the system (8)).

Results obtained show that non-linear equation system (8) has low sensitivity to errors in the initial experimental data. Identification error does not exceed 4% (for all motors power range) for all parameters of equivalent circuit (with errors in the input data not exceeding ±1%).

5 CONCLUSIONS

The use of asymmetric power supply allows to identificate the equivalent circuit parameters of the squirrel-cage induction motor when the rotor is stationary. For this purpose, DC and AC supply voltages are used. Equivalent circuit parameters are obtained as the result of solving the nonlinear equations system using numerical method.

The method does not require no-load operation mode that is very important if there is mechanical connection with the mechanism. Experimental data needed for identification are obtained by supplying two stator phase by AC and DC voltage.

Also it is important to state that:

– the assessment of the impact of errors in the initial experimental data on the identification result is important and necessary;

– estimation of influence may be performed, for example, by considering all possible combinations of error variations in the experimental data;

– degree of error values influence in the input data should be an additional criterion for evaluation of the identification method possibilities.

System (8) has low sensitivity to errors in the initial experimental data that is of high importance for modern electric drive control systems (error does not exceed 4%). Expressions presented in the paper can serve as a basis for the development of algorithms and software for microprocessor systems of AC drives with induction motors for the identification of equivalent circuit parameters.

REFERENCES

Rodkin D.I., Zdor I.E. 1998 *Modern methods of parameters estimation of induction motors after their repair* (in Russian). The problem of creating new machines and technologies: Proceedings of the Kremenchug State Polytechnic Institute, Issue 1: 100–106.

Rodkin D.I., Romashikhin Yu.V. 2009. *Equivalent circuit of induction motor in problems of identification* (in Russian). Electroinform. Problems of automated electric. Theory and practice: 392–396.

Kalinov A.P. & Cherniy A.P. 2004. *Rational approaches for determination of the parameters of AC machines* (in Russian). Bulletin of the Kremenchug State Polytechnic University, Issue 2: 115–118.

Peixoto Z. & Seixas P. 1999. *Parameters Identification for Induction Machines at Standstill.* Proceedings of the 8-th European Conference on Power Electronics and Application, CD-ROM.

Conte R.N., Pereira L.F.A., Haffner J.F., Scharlau C.C., Campestrini L., Fehlberg R.P. 2003. *Parameters Identification of Induction Machines Based on Frequency Response and Optimization Techniques.* Proceedings of the 29-th Annual Conference of the IEEE Industrial Electronics Society, CD-ROM.

Ivanov-Smolenskiy A.V. 1980. *Electrical machinery* (in Russian). Moscow: Energy: 928.

Bessonov L.A. 1996. *Theoretical fundamentals of electrical engineering: electrical circuits* (in Russian). Moscow: Graduate School: 638.

Beshta O.S. & Semin A.A. 2012. *Special feature of using of T-shaped induction motor equivalent circuit for problems of parameters identification* (in Russian). Electromechanical and Energy Systems. Problems automated electric: Theory and practice, Issue 3 (19): 553-555.

Danilov A. 2004. *Modern industrial current sensors* (in Russian). Modern electronics, Issue 10: 26–35.

Energy indexes of modern skip lifting plants of coal mines

Yu. Razumniy & A. Rukhlov
State Higher Educational Institution "National Mining University", Dnipropetrovs'k, Ukraine

ABSTRACT: The article deals with the actual power indexes of skip lifting plants with asynchronous controlled-velocity electric drive. The "technological" efficiency of such systems is shown. The main advantages and disadvantages of transformers application for the coal hoisting are given. The necessity of use of the controlled filter-compensating devices for decision of power quality and reactive-power compensation problems is grounded.

1 INTRODUCTION

The mine lifting machines behave toward plants of cyclic action, the operating mode of which is characterized alongside consistently repetitive cycles. Every work cycle is alternation of unsteady (starting, acceleration, braking) and steady (motion with the permanent speed) operating modes.

Directly the quantity of lifting cycles in day relies on many factors, basic from which are depth of the served level, carrying capacity of skip and diagram of its motion speed for the lifting cycle (tachogram), which most complete shows the kinematics mode of lifting system. Type of realized tachogram in every case relies on the row of features of lifting plant (LP) and, above all things, from the technological scheme, vessel type, method of its unloading and electric drive system of lifting machine.

2 AIM OF THE WORK

Aim of the work – establishment of conditions of the energy efficiency use of skip lifting plants with controlled electric drive systems.

3 ACCOUNT OF BASIC MATERIAL

The LP's operating mode, at which lifting tachogram by form will approach to rectangle (to have maximally possible acceleration and deceleration), will be characterized by the minimum consumption of electric power. This is achieved by application of the modern electric drive systems on the base of thyristor converter which allow effectively to regulate acceleration and deceleration in the unsteady periods of lifting tachogram (with the observance of their legitimate values resulted in the normative documents, such as Safety Rules, Rules of Technical Exploitation and other). This is the basic difference of such systems from the "traditional" relay-contact scheme with lifting rate control by the active resistances in the rotor circuit of asynchronous motor (AM).

From the power point of view the basic dignity of application of the controlled electric drive system there is an economy of electric power, the size of which can arrive at 10–20 % in comparison with application of the rheostat rate control of lifting. In addition, such systems have row of other "technological" advantages (Razumniy, Rukhlov & Kramarenko 2014):

- the rise of safety, reliability and continuity of LP operating, that will secure the normal rhythmic work of enterprise;

- the more exact and smooth speed regulation of electric motor allows to give up the use of reducing gears and other additional apparatus, that considerably simplifies a mechanical (technological) scheme, promotes its reliability and lowers running expenses;

- the controlled starting of the guided electric motor secures its smooth acceleration (without the promoted starting currents and mechanical blows), that lowers loading on motor and, accordingly, multiplies a term of its exploitation;

- the rise of LP productivity on 10–15 % due to the "compression" and adherence of set lifting tachogram, self-controls of pauses between cycles at the load-unloading of skips, automations of auxiliary operations and other;

- the visual control of generous amount of the technological and electric parameters of lifting plant (for example, skip position in shaft, quantity of lifting cycles for days, change of values of supply voltage and current).

To the basic disadvantages of application of the controlled electric drive systems on the base of thyristor converter belong:

- generation of considerable electromagnetic interferences which show up in the harmonicity distortion of supply voltage and current curves;

- low value of power factor especially at the deep rate control of lifting, that shows up in the considerable consumption of reactive power by LP;

- possible reduction of efficiency coefficient and service life of motor, additional losses of power and energy related to worsening of electric power quality;

- considerable capital costs and other.

The presence of substantial advantages of application of energy transformers stipulated their wide introduction in the electric drive systems of different technological processes, including for skip LP of the coal mines of Ukraine. On Figure 1 actual day's electric loading diagram (ELD) of the skip lifting plant with the controlled electric drive system "asynchronous thyristor cascade" (ATC) on base AM by power 800 kVt is shown.

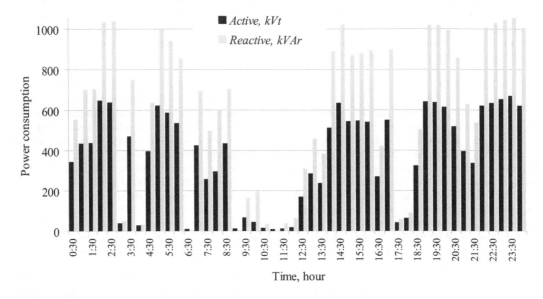

Figure 1. Actual day's ELD of skip LP with asynchronous electric drive by power 800 kVt with the ATC system

For electric drive by ATC system connecting to the power grid 6 kV is characteristic for stator (straight) and rotor (through the special transformer). Therefore on Fig. 1 the diagrams of active and reactive powers totally for AM's stator and rotor are resulted. In addition, on ELD the volume of generated electric power due to its recuperation in the electric network (that characteristically for the cascade scheme, such as ATC) is not shown, because the quantity of the generated active power does not exceed 1–2 % from the consumable value.

Data on Figure 1 testify to the substantial unevenness of the ELD of coal skip LP (average value of form coefficient $k_f = 1.2 - 1.25$), which is determined by a different quantity of lifting cycles for the half-hourly average interval (the standard deviation arrives at 30 % from the nominal motor power); characteristic low level of electro-

consumption in the repair change, caused by absence of the transported coal from interruption under the authority of production works; exceeding of reactive power consumption on comparison from active one, which determines the low average value of power factor at level 0.4 – 0.55 (one of the basic disadvantages of the controlled electric drive systems on the base of thyristor converters).

The fragments of measuring as curves of change of energy indexes in the characteristic work cycle of skip LP with asynchronous electric drive by ATC system will help to confirm the last failing evidently. Its resulted on Fig. 2. The forms of the resulted curves testify to the considerable consumption of reactive power in the periods of unsteady skip motion (in the process of its acceleration, braking and unloading). In these tachogram periods the reactive electro-consumption there are more of active in 2–2,5 times (Fig. 2, a). In

48

the process of steady skip motion the consumption surge of the reactive loading is not observed, but its value is commensurable with active one. The average value of power factor for the lifting cycle makes 0.4–0.45, that impermissible low (Fig. 2, b). Except for this, only in the periods of the unsteady skip motion (at lifting control rate) the negative influence of transformers on the power supply system, which consists in the generation of the higher level of voltage and current high harmonic, shows up. As a result the voltage nonsinusoidality ratio can exceed the legitimate values (Fig. 2, c). The resulted fragments of change of actual energy indexes confirm that the "deep" adjusting of the LP's technological parameters by thyristor converters considerably lowers a power-factor and increases nonsinusoidality of supply voltage.

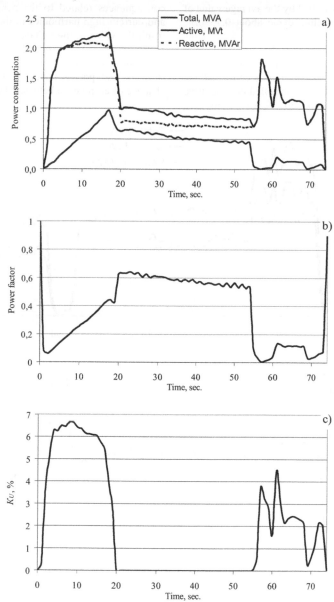

Figure 2. Changing of the energy indexes in lifting cycle with asynchronous electric drive by ATC system

The changing curves of energy indexes for the actual lifting cycle of the same skip LP, the parameters of which are shown on Fig. 1 and 2, only during its work on the relay-contact scheme of control with the active resistances in the rotor circuit, are resulted on Fig. 3. The forms of the resulted curves cardinally differ from the Fig. 2. At first, the value of reactive power consumption comparably with active one for all periods of tachogram, that is multiplied by the average value of power factor for the lifting cycle about 0.75–0.77 (Fig. 3, a, b).

Secondly, duration of skip motion increases on 3 seconds (from 74 to 77 sec.) on comparison from electric drive by ATC system, that is confirmed the advantage of application of controlled electric drive in part of rise of LP's productivity. Except for this, the active electro-consumption increases on 11.9 % on comparison from the ATC system due to the insignificant increase of average power (from 403.8 to 440.5 kVt) and duration of lifting cycle. For the given skip plant the introduction of the controlled electric drive system allowed to advance a lifting quantity for day by 20–30 cycles. Thirdly, the absence of thyristor converters in the relay-contact scheme of control allows to evade the negative consequences related to the generation of voltage and current high harmonics in the power grid. As a result the value of the voltage nonsinusoidality ratio K_U exceeds 0.1 %, but the curve form of its changing does not rely on the form of other indexes (powers and power factor), that is confirmed by absence of correlation between them (Fig. 3, c).

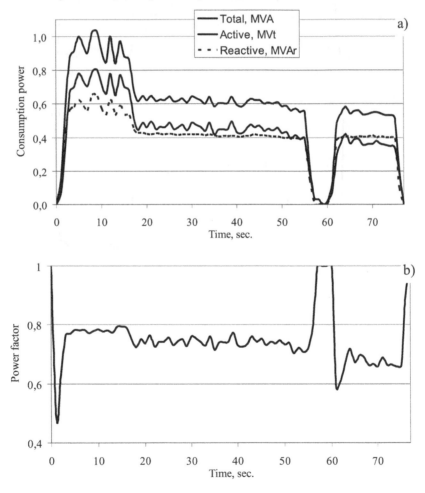

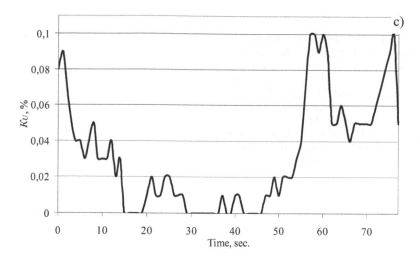

Figure 3. Changing of the energy indexes in lifting cycle with asynchronous electric drive and relay-contact scheme of control

The comparison of descriptions on the Fig. 2 and 3 confirms two basic advantages of application of the ATC system (decrease of electric power consumption and rise of LP's productivity) and disadvantages (considerable lowering of power factor and generation of voltage and current high harmonics). The installation of the special devices for reactive power compensation and high harmonics filtration for minimization of negative consequences in the power supply system is needed.

At voltage and current nonsinusoidality in the power grid the reactive power compensation for increase of power factor by the most widespread static capacitor bank is considerably complicated or generally turns out impossible. This is related to a few principal reasons. At first, the capacitor banks can long-term work at the overload by the high harmonic currents no more than on 30 %, and the possible increase of voltage on their clamps makes 10 %. However the resistance of exactly condensers considerably diminishes with the increase of frequency, therefore its service life grows short in such conditions (especially for batteries which are connected to the grid directly, that is without the protective reactors). Secondly, the capacity value of condensers and inductance of power grid can provoke in the power supply system mode near to currents resonance on frequency of any from harmonics. Of course, such mode results in the overload of capacitor banks and return them from line-up (Zhezhelenko, Shidlovskiy, Pivnyak & Sayenko 2009). Thirdly, the abrupt variable mode of reactive power consumption (brief surge and dips – see Figure 2, a) at application of the electric drive systems on the base of thyristor converters does

ineffective the use of not controlled or stepwise controlled capacitor banks. For such conditions the introduction of filter-compensating devices is needed, the quantity of the generated reactive power of which changes in real time. The controlled filter-compensating device secures the nonsinusoidality decrease of voltage and current curves due to filtration of the proper harmonics, and also supports constantly the high value of power factor (near to 1) on the substation buses to which it is connected. It is necessary to mark that there is positive experience of introduction of such devices on the coal mines of Ukraine.

4 CONCLUSIONS

1. The operating modes of skip lifting plants with the controlled electric drive systems on the base of thyristor converters are characterized of low power factor values (0.4–0.45) and high levels of voltage nonsinusoidality ratio (K_U arrives at 8 % and more).

2. The problem of increase of the electric power quality in the power grids with the nonlinear loading, which is intensified lately from the wide introduction of the electric drive systems on the base of thyristor converters, needs the complex decision both in part of decrease of nonsinusoidality and in part of reactive power compensation. The one of effective methods of its decision is the application of the controlled filter-compensating devices in the power supply system.

51

REFERENCES

Razumniy Yu.T., Rukhlov A.V., Kramarenko S.A. 2014. *Improvement of energy efficiency of the main dewatering plant of a coal mine* (in Russian). Dnipropetrovs'k: State Higher Educational Institution "National Mining University", Mining Electromechanics and Automation, Issue. 92: 7–11.

Zhezhelenko I.V., Shidlovskiy A.K., Pivnyak G.G., Sayenko Y.L. 2009. *Electromagnetic compatibility of electrical supply* (in Russian). Dnipropetrovs'k: National Mining University: 319.

Power Engineering, Control and Information Technologies in Geotechnical Systems – Pivnyak, Beshta & Alekseyev (eds)
© 2015 Taylor & Francis Group, London, ISBN 978-1-138-02804-3

Informational and methodological support for energy efficiency control

N. Dreshpak & S. Vypanasenko
State Higher Educational Institution "National Mining University", Dnipropetrovs'k, Ukraine

ABSTRACT: Energy efficiency control validity on the basis of regression analysis at industrial enterprises is confirmed. The factors that determine control frequency and constant plan indicators duration are defined. A criterion that determines regression dependences substitution order is proposed.

1 INTRODUCTION

Efficient energy use is estimated by specific consumption calculating, which is energy consumption per unit of manufactured products. This improvement has become possible due to efficient energy control at industrial enterprises (Shidlovskii, Pivnyak, Rogoza & Vypanasenko 2007).

Energy efficiency control involves comparing the actual parameter with its plan indicator. The results of this comparison allow us to make conclusions about effective or ineffective enterprise performance, technological lines etc. These results depend on management. In case of satisfactory energy efficiency indicators of the process it is necessary to pay bonuses to the stuff in order to encourage their efficient energy use. Otherwise the staff performance in energy efficiency will be considered as unsatisfactory. Thus, this control is very important for staff performance rating and therefore the results are deeply analysed in terms of their validity. It is very important because energy control validity degree determinesperformance "fairness." If workers don't understand the reasons of control based decisions, we should not rely on their energy saving support.

The task of this publication is to estimate the validity and identify characteristics of the control procedure which is based on regression analysis.

2 FACTORS AFFECTING CONTROL RESULTS VALIDITY

Let us examine the factors that affect control results validity. The procedure involves two main components: the actual and energy efficiency plan indicators. To determine the actual energy efficiency performance we measure consumed energy by using devices that provide required accuracy. Thus, having accurate data on production output the accuracy of energy efficiency performance depends on the accuracy of measuring devices. The task of setting plan indicators, which consider production factors, is the most complex. These targets can be set on the base of forecasting energy consumption data and that is why the forecast's accuracy is so important. The forecast is based on past production output analysis, that is why regression analysis is one of the most efficient methods of forecasting. An approach, which defines plan indicators in the form of specific energy consumption standards, is widely used in practice.These standards are used both - in specific industries and in industrial enterprise's production lines. Specific industry standards have disadvantages – they don't take into account each enterprise's production peculiarities. In this regard, for some enterprises these specific consumption standards are too high and for others - too low. Sometimes the standards define an acceptable range for variation of established parameter or its upper limit. It is clear that in this case the target figure is "vague" and control effectiveness is reduced because the indicators are not substantiated enough.

Approved methods are used to calculate specific consumption standards of industrial enterprise's production lines, which partially take into account production process peculiarities. At the same time, simple analytic dependences are used and they do not provide accurate plan indicators. Moreover, these standards are valid for a long period which sometimes could be several years. Significant changes in the structure of an enterprise and its equipment can be in this period making these standards impossible to use.

In a view of these disadvantages we can form some important principles to be followed in order to ensure these indicators validity:

-these plan indicators must take into account production process peculiarities of the enterprise;

-these plan indicators must take into account energy consumption changes that occur in energy consumption of the enterprise during the control period.

3 REGRESSION ANALYSIS FOR ENERGY EFFICIENCY CONTROL

Regression analysis could be used to form the plan indicators according to the principles mentioned above. Regression analysis involves gathering information to construct dependence of parameters that directly characterize the production process at a certain enterprise. Moreover, such a gathering and further data processing can be updated often enough, providing plan indicators adjustment with regard to production processes changes. Regression analysis as a method of mathematical statistics allows parameters processes dependences reproducing in a simple analytical form. In this regard a great variety of variable data (multiple regression) as well as existing dependences (linear or non-linear regression) can be taken into account. Simple analytical dependencies are easy to use, especially when we constantly update their characteristics (while controlling). Representing these dependencies in two- or three-dimensional images we can see that they are very visible and informative. These images represent the dependence between energy costs and production output, and this feature is attractive while analysing energy efficiency of a technological process. These dependences are constructed for different control intervals and allow dynamics change control at all stages. Taking into account the importance of energy efficiency indicatorssuch information is urgently required for any energy-intensive industries because they show energy saving positive and negative tendencies.

In case of multiple regression using for energy consumption level determining, we should simultaneously record the values of all the parameters influencing this level. In terms of existing production it is not so easy to solve the problem. Parameter list increasing requires additional efforts to gather information. At the same time, computer accounting systems usage is not always meet the targets because of their limited performance as well as their high value. Therefore simple dependences are widely used in energy management systems with limited number of parameters to be measured. Their number is no more than three usually. A dependence of energy loss(E) on one parameter-production output(P) is widely used. It is important that E (P) dependence is in full compliance with energy efficiency plan indicators, allowing rather easy to get E /P (P) value – the dependence between specific energy consumption E / P and production output (P).

Note: a type of dependence E (R) (linear or nonlinear) is important. Based on the principle of regression dependence (least - squares method) a type of dependence can be matched according to minimum value of dispersion providing which depends on experimental points coordinate deviation from proposed curve. At the same time, in cases when dispersions values of different dependences are slightly different, we recommend to use linear regression - the simplest form (E = A + BP, in which A, B - regression line coefficients). This linear regression uses parameters which common in energy management: A, BP- constant and variable components of energy consumption, correlation (r) and determination (r^2) coefficients (Vypanasenko & Dreshpak 2013). Thus, its use allows us to analyse the plan indicators which is important for current production, providing constant analysis of indicators and their dynamics. It is known that E (R) linear dependence thirst of all should be applied in conditions of "uniform accumulation law" acting and the limits of P-parameter changing are minor (Vypanasenko 2008).

Let us focus on plan indicators validity and accuracy. Notice, that having limited data while regression line construction this dependence is random. A and B regression coefficients are calculated with certain inaccuracies. Therefore it is necessary to form a confidence interval in which it is a high probability of true (correct) regression dependence (Fig. 1). You can see that a confidence interval (a shaded part) is bounded by four straight lines matching the values of possible deviations of A and B coefficients ($A \pm \delta'$, $B \pm \delta''$). The dependences (Vypanasenko 2008) should be used to calculate these values.

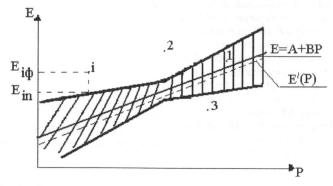

Figure. 1. Regression line location

Computer programs to calculate A and Bcoefficients of linear regression allow getting the values root mean square deviations δ' and δ'', as well as defining the limits of the line's location. A confidence interval in energy efficiency plan indicators determining slightly reduces control accuracy. Indeed, if the actual value coordinates of energy consumption (e.g.point1) fit the shaded part (when compared with plan indicators), matching result can be deemed as "satisfactory pertaining to accuracy control". A location of points outside the shaded areas means a high probability of energy consumption average value exceeding (e.g. point2) regarding to certain production output, or have a lower level (e.g.point3 which describes good energy efficiency performance). It is clear the desire of control procedure executor to reduce the shaded area for the purpose of control accuracy increasing which allows us to make more conclusions about "good" or "unsatisfactory" results of efficient energy use. For this purpose it is necessary to enhance dependence between E and P parameters (to increase correlation coefficient), which can be achieved by other factors influence reducing on the rate of energy consumption (Vypanasenko & Dreshpak 2013).

4 ENERGY EFFICIENCY CONTROL PROCESS

Ongoing control- based management allows us to improve efficient energy. These changes can be both structural (machinery changing) and operational (operating mode of machines and mechanisms changing). Mode operating should be used first of all because it does not require big capital investments (we do not recommend to use low efficiency equipment, we recommend technological process automation usage, etc). These changes lead to energy indicators adjusting. This can be done through a new regressive dependence using basic data obtained at production sight just before a control procedure. Fig. 2 shows a control process which involves plan indicators periodical changes (regression dependences). It is clear that the whole control period is divided into intervals involving actual and plan indicators comparison, as well as basic data gathering (actual parameters) for next interval indicator ssetting (new regression line construction).

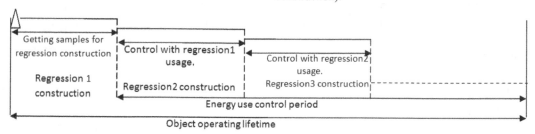

Figure. 2. Energy efficiency control process with plan indicators periodic changes

The initial phase which is associated with regression1 construction probably will not allow getting the best plan indicators results because initial energy consumption is significant. A control procedure will improve the situation reducing energy consumption over a period of control. Thus, energy efficiency improvement provides us step-by-step progress to a higher value of this parameter. It is possible only in terms of production systematic control. Optimal process development could be deemed when each successive $E'(P)$ regression line is below the previous $E(P)$ regression line (Fig. 1). In this case, each successive control interval is characterized by more high-value plan indicators. If there are no structural changes at production sight, the regression line will be located in the way providing maximum energy efficiency.

The process development may be different from optimal. In some control intervals most of the actual energy consumption indicators could exceed plan indicators and therefore you can layout the following regression line above the previous one. This variant is not acceptable because the previous levels of energy efficiency demands are reduced. We suggest not substituting one regression dependence with another in the next control interval, the previous one should act. This decisions reasonable in case of there are no more important production factors influencing energy consumption growth (e.g. job safety).

5 CRITERION FOR REGRESSION DEPENDENCES SUBSTITUTION

We propose to define a criterion which allows determining a feasibility of one regression line substitution with another. It refers to cumulative sum value (Vypanasenko 2008). An actual energy consumption value and plan value deviation is calculated using (energy savings (+) and excess consumption - (-)) signs in control interval. For example, for the i-st observation (Fig. 1) the deviation of actual value E_{ia} from the plan E_{ip} is negative

$$K_i = -\left(E_{ia} - E_{ip}\right) \tag{1}$$

and then cumulative sum

$$K = \sum_{i=1}^{n} K_i, \tag{2}$$

where n – interval control number of observations.

Note that confidence interval limit values for line regression are suggested for energy consumption plan indicators(E_{ip}). This provides accurate information about energy excess usage or saving.

If a cumulative sum K has + sign in control interval (n – observations), it shows a good staff performance, and these achievements should be taken into account when setting the indicators for the next interval. It makes sense to substitute one regression line with another. If we have negative K value substitution should not be used. In this case a preceding regression line should be used for energy use control in the next interval. This makes the stuff to achieve these results in the future.

As an example, we represent you the sequence of cumulative sum K calculating during energy use control at one of the coal mines of Ukraine's Western Donbas. We recorded coal output and the corresponding values of power consumption every day during the month. This enabled us to construct a linear regression dependence $E(P)$ on the base of experimental data and define a confidence interval for this dependence. We controlled energy use at coal mine during the next month. We defined deviation of actual energy consumption from plan indicators(K_i) every day. Table 1 shows daily (n=30) deviations with regard to their signs.

Table 1. Daily K_i values deviations

Day of month (i)	1	2	3	4	5	6	7	8	9	10	11	12	13	14	15
Values K_i, kilowatt-hours	1500	-1100	750	1100	-2000	720	-1140	1100	1840	-680	-1820	920	-400	1200	980

Table 1 continuation

Day of month (i)	16	17	18	19	20	21	22	23	24	25	26	27	28	29	30
Values K_i, kilowatt-hours	-1500	2010	-840	-30	200	820	760	-300	-1000	480	1200	-1000	-840	670	500

According to (2) the cumulative sum is 4100 kilowatt-hours showing us a good stuff performance in energy saving. A positive result must be considered while setting the plan indicators for the next control period (a month). This means you should substitute one regression line with more effective one. If the dynamics of plan indicators substituting is positive, it shows us about successful energy management at the enterprise. In case when the changes do not occur or too slow compared to the initial plan indicators, it is necessary to apply more effective measures.

6 FREQUENCY AND DURATIONOF ENERGY EFFICIENCYCONTROL

Control frequency should depend on deviation of actual energy consumption from plan indicators. In case of significant deviations you should react urgently without delays. It is also necessary to inform the workers about the situation on production sight. Moreover, control frequency depends on existing ability to obtain an accurate information about actual energy consumption and production output. It is mainly determined by automated energy measurement systems existence or absence. For example, at many mines in Donbas such systems are installed, which allow energy consumption daily values recording and comparing them with coal daily production. This provides an opportunity for coal mining efficiency daily control.

This control duration depends on energy equipment's structural and operational changes. It is obvious that these changes should be taken into account in plan indicators and while new regression dependence constructing. Control-based energy consumption management also leads to equipment' operational changes (overloads and no-load operation elimination, operation time optimization). Feedback action improves the situation on sight and also requires plan indicators changing. It is also necessary to remember that new data gathering for new regression line occurs during the control period. An amount of 'tests' must be sufficient for regression line construction with rather narrow confidence interval (for the purpose of control accuracy). Thus, the analysis shows that having daily energy consumption control at a coal mine, it is reasonable to limit the control interval to one month. Getting thirty results of "testing" for regression line construction provides energy consumption control acceptable accuracy. The necessity of new regression line construction can be confirmed by significant positive cumulative sum of K deviations obtained at the end of control interval. The K value is an integral parameter that characterizes the difference between actual results and plan indicators.

Therefore, it is possible to connect the K value with bonuses for the stuff for energy saving. For example, if it is one - month control period and a positive K value has been obtained, the stuff deserves bonuses which are proportional to the K value.

7 CONCLUSIONS

Energy efficiency control frequency should be connected with both - deviations of actual energy consumption from plan indicators and existing facilities for gathering reliable information to realize the control. Control interval duration depends on energy equipment's structural and operational changes that occur at controlled sight.Cumulative sum is a criterion that determines the order of regression dependences substitution.

REFERENCES

Shidlovskii A., Pivnyak G., Rogoza M. & Vypanasenko S. 2007. *Geopolitics and geo-economics of Ukraine* (in Ukrainian). Dnipropetrovs'k: National Mining University: 282.

Vypanasenko S. & Dreshpak N. 2013. *The features of energy efficiency measurement and control of production processes.* Energy Efficiency Improvement of Geotechnical Systems - Proceedings of the International Forum on Energy Efficiency. Netherlands: CRC Press/Balkema: 71-78.

Vypanasenko S. 2008. *Energy management systems of coal mines* (in Ukrainian). Dnipropetrovs'k: National Mining University: 106.

Implementation of the insulation resistance control method for high-voltage grids of coal mines

F. Shkrabets & A. Ostapchuk

State Higher Educational Institution "National Mining University", Dnipropetrovs'k, Ukraine

ABSTRACT: The article presents a new principle of the insulation resistance control device operation of high-voltage grids of mining enterprises under operating voltage, which is based on the superimposition of bi-frequency operational sinusoidal signals of nonindustrial frequency on the distribution grid, which can significantly increase the sensitivity of these devices and make them almost independent of the total ground capacity of the grid. Development of these protection devises operating principles allow significantly increase grids' reliability and electrical safety.

1 INTRODUCTION

Electrical power systems grids and equipment of mines and quarries in Ukraine are operated in very specific circumstances, which to a large extent determine the level of power supply reliability and electrical safety. Improving of the electrical safety conditions and the reliability of mining facilities power supply depends on the successful resolution of the range of issues, the most important of them are problems of creation of methods and tools that provide prevention, search and restoration of the power supply damages. Analysis of operational crash data in mining facilities distribution grids shows that the share of ground faults is up to 65…90% of the total number of failures, most of which are related to the phase-to-ground insulation violation of the grid, i.e. to asymmetric faults emersion /1, 2/.

2 FORMULATION THE PROBLEM

Most of the damages in the distribution grids lead to a decrease in the level of electrical safety and reliability of electricity supply. Further, the reduction of power supply reliability leads to an increase in the share of losses from power supply outage. In general, this damage is determined by the duration of mining and transport machines idle-time and, thereafter, leads to product undersupply. Unwarranted downtime of mining and support mechanisms occurs as a result of wrong action of first and second protection level devices against ground faults due to the significant time searching for missing damages (with a false protection devices actuation). In addition, the ground-faults are often the cause of multi-

phase damage, thereby increasing the amount of machinery idle.

3 MATERIALS FOR RESEARCH

The reliability of power supply is largely dependent on the quality of the ground fault protection functioning in the distribution grid. Furthermore, it was found that, ceteris paribus, the quality of named protections depends on the structure and construction of the neutral mode. The studies have shown that the main cause of false protection actuation against ground faults in networks with fully insulated or compensated neutral should be considered as the emergence in the network, after turning off the damaged connection (or damage self-destruct), an oscillatory process with a frequency close to the 50 Hz. In general, the time-changing process of the neutral shift voltage in system after disconnection or damage disengagement can be described by a differential equation

$$\frac{d^2 U_0(t)}{dt^2} + \left(\frac{3R_r + R}{3\omega C R_r + R} \cdot \frac{dU_0(t)}{dt} \right) -$$

$$- \frac{1}{3CL_p} U_0(t) = 0$$

whose solution and results analysis is made with a real parameters of distribution grids with different types of ground, led to the following conclusions:

1. In power networks with completely insulated neutral the transient process form (natural frequency and duration of the process) is mainly determined by the total grid capacity relative to the ground and the number of simultaneously switching voltage

transformers. For real distribution grids parameters, the transient duration ranges from 2 to 10 commercial frequency cycles, and natural frequency value is typically less than commercial frequency, and the free oscillations frequency, directly in the damping process, changes due to nonlinear character of measuring voltage transformers reactance.

2. In compensated neutral networks the voltage damping is determined mainly by relative to the ground insulation parameters of the distribution grid and doesn't depend on the compensation device params. Natural frequency and the time constant of decay in compensated networks significantly higher than similar characteristics in networks with a fully insulated neutral. The nature of the transient process, ceteris paribus, depends on the compensating device settings.

3. The transient process in networks with the neutral resistor to a large extent dependents on the value of the resistor. If resistor is installed with resistance, chosen from the condition that active component of the single-phase ground fault current value is 50% of the capacitive current, due to a sharp increase in the damping coefficient, the transient process is almost finished in a half-period of industrial frequency.

To study the characteristics of zero-sequence voltage and current in the steady ground fault state we used the equivalent circuit of the two connections that are connected to a single power transformer. In this case, the admittances of the applicable phases insulation of controlled connection ($Y_{A1} = Y_{B1} = Y_{C1} = Y_1$) and all the rest of the distribution grid ($Y'_A = Y'_B = Y'_C = Y'$) are related by the ratio $Y_1 + Y' = Y$. For this scheme, in general, to the condition of single-phase ground fault in a controlled affixion or in the external network, could be written an expression:

- for the zero sequence voltage

$$\underline{U}_0 = -\underline{U}_\phi \frac{y_1}{3Y + Y_H + y_1}; \text{ or}$$

$$\underline{U}_0 = -\underline{U}_\phi \frac{y}{3Y + Y_H + y},$$

- for the zero sequence current at the damage in controlled line

$$\underline{I}_0 = -\underline{U}_0(3Y + Y'_H) =$$

$$= -\underline{U}_\phi (3Y - 3Y_1 + Y_H) \frac{y_1}{3Y + Y_H + y_1}$$

- for the natural current in controlled line (zero sequence current at the external ground fault in controlled line)

$$\underline{I}_{0C} = 3\underline{U}_0 Y_1 = -3\underline{U}_\phi \frac{y}{3Y + Y_H + y},$$

where Y_H – neutral point admittance relative to the ground.

The studies of these relations on the impact on the zero-sequence voltage and current values of the insulation and the neutral operational mode and their analyze lead to the conclusion:

- zero-sequence voltage is determined by the parameters of network insulation relative to the ground, by network neutral point admittance relative to the ground (the degree of arc suppression coil resonant mode mismatch (for compensated grids) and the transient resistance value at the point of phase-to-ground.

- in a phase-to-ground short circuit (zero impedance) the compensating device mode does not affect on the value of zero-sequence voltage, which at such damages equals to a phase voltage of the power network;

- zero sequence current in the damaged line is defined by zero sequence voltage, by insulation parameters of an external network relative to the ground, that is, parameters of insulation of all network relative to the ground (including neutral point admittance) minus parameters of insulation of the damaged affixion;

- natural current of controlled affixion (zero sequence current at the external damage in controlled line) is determined by zero-sequence voltage and insulation parameters relative to the ground only in controlled affixion;

- current in the network neutral (in the compensating device or in the neutral resistor) at ground fault is defined by zero sequence voltage and by the compensating device or the resistor direct parameters.

For directional protection devices responsive to the zero-sequence power is also required to consider the phase or the mutual position of the compared values, i.e. the voltage and zero sequence currents vectors position. The characteristics of natural current of the protected line and current in a network neutral are of interest from the point of the creation of new methods and tools of directional protective devices or ground fault alarms

By phase characteristics research results taking into account real values of parameters of insulation relative to ground of the whole network and separate affixion, and also in view of a real mutual ratio of capacitor and active resistance of insulation, it is possible to draw the following conclusions:

1. The Grid with the full insulated neutral

- the angle between the zero sequence voltage vector and the damaged phase voltage vector changes ranging from 180 to 90 at change of transient resistance in a short circuit point from zero indefinitely;

- the angle between the zero sequence current vector and the zero sequence voltage vector doesn't depend on completeness of short circuit (transitional resistance in a short circuit point) and makes nearly 270 el. degrees, or minus 90 el. degrees.

- the angle between the own current vector of the controlled line (zero sequence current in the controlled line at external short circuit of one phase) and the zero sequence voltage vector is defined by relative to the ground insulation parameters of only controlled affixion and makes nearly 90 el. degrees.

2. The Grid with the compensated neutral

- the angle between the zero sequence voltage vector and the damaged phase voltage vector depends on parameters of network insulation, parameters of the compensating device and transient resistance value in a place of short circuit and can accept values in the range from 90 to 270 el. degrees, i.e. theoretically within 180 el. degrees. When the compensating device is tuned-up in a resonance with a network capacity relative to the ground the specified angle is almost equal 180 el. degrees and doesn't depend on parameters of a network and transient resistance value in a short circuit point;

- the angle between the zero sequence current vector and the zero sequence voltage vector doesn't depend on completeness of short circuit (transient resistance in a short circuit point) and is defined substantially by value of compensating device detuning from a resonant mode and for resonant tuning makes almost 180 el. degrees; in case of compensating device detuning from a resonant mode, both towards overcompensation, and towards undercompensation, the current vector deviates on an angle, accordingly, to plus and a minus 90 el. degrees, in such way the range of angle change theoretically makes 180 el. degrees;

- the angle between the natural current vector of the controlled line and the zero sequence voltage vector is defined by insulation parameters related to ground of only controlled affixion and makes 90 el. degrees;

- the angle between the current vector in the arc suppression coil and the zero sequence voltage vector is determined only by the arc suppression coil parameters and for their real value is 90 el. degrees.

3. The Grid with the resistor in a neutral

- the angle between the zero sequence current vector and the zero sequence voltage vector doesn't depend on completeness of short circuit and, unlike networks with full insulated neutral, is equal to value ranging from 180 to 270 el. degrees; for real parameters of network insulation related to the ground, and the recommended value of $R_H = (1...2)X_C$ this angle is about 225...240 el. degrees;

- the angle between the natural current vector of the controlled line (zero sequence current in the controlled line at external short one phase circuit) and the zero sequence voltage vector is determined by the parameters of the related to the ground insulation of only controlled affixion and is almost 90 el. degrees;

- the angle between a current vector in the resistor which has been switched on in a neutral of a grid and the zero sequence voltage vector doesn't depend on parameters of insulation of the network, the resistor and a short circuit mode and matches in the direction with the zero sequence voltage vector (the angle is equal 0 el. degrees).

It is important to note that the natural current phase of controlled affixion doesn't depend on a neutral operating mode, and is defined only by parameters of directly controlled affixion and is almost rigidly attached to zero sequence voltage.

The received results formed a basis for development of a way and creation of the protection device against leaks and short circuits on the ground for the mine distributive grids, using for definition the damaged affixion, except zero sequence voltage and current, the phase voltages of the grid /1/.

The principle of this protection method is based on the performance evaluation of the relative position of zero sequence current vectors and phase voltage vectors. Based on a sinusoidal signal phase voltage pulses are formed with a time interval which is equal to 120 °. The zero-sequence current signal also converted into pulses shifted in phase by 90 °, with an interval of 360 °. At coincidence pulses of voltage and current (corresponding to the damaged affixion), the signal to turn off the damaged circuit is formed. Thus on the unimpaired affixion impulses of zero sequence current are displaced on 180 ° and don't coincide with phase voltage impulses. This principle can significantly increase the sensitivity of the protection devices against ground faults and make it virtually independent of the total grid capacity, related to the ground.

Double ground short circuits in different points of a distributive grid are, as a rule, development of single-phase damages and are caused by influence of the internal overvoltages appearing thus. It is necessary to distinguish two main types of double short circuits on the ground: damage of related to the ground insulation of the same phase in two points of a distributive grid; damage of related to the ground insulation of different phases of a distributive grid.

Damage to the insulation of the two points of one phase of the distribution grid can be represented by the following failure modes: damages to two points of the controlled line; in two points of an external grid; in the controlled line and in an external grid. These emergency conditions correspond, in general, a single-phase ground short circuit via the contact resistance whose value is equivalent to two parallel resistances at different points in the network. The third case at which the zero sequence current value in the controlled line will depend on a ratio of transitional resistance value is of interest and with sufficient accuracy for practical calculations determined by the expression:

$$I_{01}^{(1)} \approx I_3^{(1)} r / (r_1 + r)$$

where $I_3^{(1)}$ – total current of single-phase short circuit on the ground.

The second type of double short circuit has several varieties, differing in the mutual arrangement of damaged areas concerning the controlled affixion. In this case it is necessary to observe the following emergency modes in a insulated neutral network:

- related to the ground insulation damage in two phases of the controlled line:

$$\underline{I}_{01}^{(1.1)} = -3\underline{U}_0^{(1.1)}(Y - Y_1), \text{ or}$$

$$\underline{I}_{01}^{(1.1)} = \frac{-3\underline{U}_A(Y - Y_1)\left(y_{1A} + a^2 y_{1B}\right)}{\left(3Y + y_{1A} + y_{1B}\right)};$$

- insulation damage to two phases to an external network

$$\underline{I}_{0C}^{(1.1)} = 3\underline{U}_0^{(1.1)} Y_1 = \frac{-3\underline{U}_A Y_1\left(y'_A + a^2 y'_B\right)}{\left(3Y + y'_A + y'_B\right)};$$

- insulation damage in the two phases at different affixions (one - damaged in a controlled phase line, the other - on the external network)

$$\underline{I}_{02}^{(1.1)} = \underline{U}_A \times$$

$$\times \frac{y_{1A}\left[3(Y - Y_1) + y'_B\right] - a^2 y'_B(3Y_1 + y_{1A})}{3Y + y_{1A} + y'_B}.$$

The analysis of research results of zero sequence voltage and currents at double short circuits on the ground in distributive grids with the insulated neutral allowed to estimate influence of distributive grid insulation parameters, contact resistance in short circuit points on character of values change and the provision of zero sequence voltage and currents vectors. It is established that value of contact resistance in damage points has defining impact on operating value of zero sequence voltage. The influence of the distribution network insulation parameters to the value of zero sequence voltage is shown mainly at relatively high values of resistances. The phase of zero sequence voltage also substantially is influenced by a mutual ratio of values of contact resistance in points of damage and to a lesser extent by parameters of a distributive grid. The position of zero sequence voltage vector in relation to the phase voltage vectors in the double ground faults can theoretically vary between 210 °. In grids with compensated neutral the angle between zero sequence voltage and current in the double ground fault is not fixed, and is determined generally by the compensating device setup mode and the contact resistance values ratio in points of short circuit and can change within 300 el. deg.

So far the automatic reclosing devices almost didn't expand in distributive grids of opencast mining. Development and deployment of such devices restrains many factors, basic of which are:

- the design of power switching distributing devices, precluding the possibility of connecting the operational power supply;

- the existence of a sufficiently long run-out voltage at the powerful quarry electro receivers;

- rather high probability of non-actuating (in consequence of the mechanical system damage) of one or all phases by the switch after command for shutdown.

The main characteristics, reflecting conditions of electrical safety at operation of automatic reclosing devices in quarry power networks and demanding justification, are:

- setpointing of automatic reclosing device ban on level of active resistance of insulation related to the ground of the disconnected affixions;

- value of the operational power supply voltage for control of the disconnected affixions insulation.

From the insulation condition assessment point of view, operational voltage for insulation control of the disconnected affixion has to be close or equal to the working voltage of a grid. However, proceeding from electrical safety conditions, providing demands are made to a source and level of operational voltage in the following conditions:

1) the source is connected to the line which current carrying parts may be touched by people;

2) the person touches already disconnected line which own capacity is loaded up to the voltage of the operational power supply.

According to the requirements of the automatic reclosing devices for quarry power networks the algorithm of work of its logical part which realizes the following functional instructions is developed:

- AR start from the action of protection against ground fault or overcurrent protection (if necessary);
- counting the time delay required to reduce to a safe level of EMF motor runout;

- control of a voltage applied to affixion of the disconnected phase and the development of an appropriate command to continue or ban reclosure;
- connect to affixion of the disconnected phase source of DC;
- related to the ground insulation resistance measurement of disconnected affixion and compare it with the set point;

The main element of the automatic reclosing devise is program block, based on the Read-only memory (ROM) and implements a given algorithm (Fig. 1).

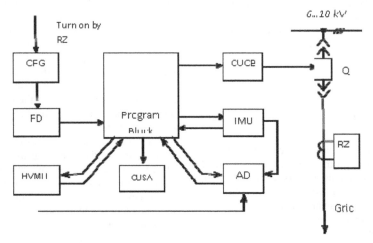

Figure 1. The functional diagram of the AR for quarry power network

As indicated in the diagram: CPG - current pulse generator, produces a square wave with the frequency of 50 Hz and necessary to synchronize the device; FD - frequency divider, designed to convert CPG to communicate with the address inputs program block; HVMU- high voltage monitoring unit; CUSA - the control unit switches of BB-20 affixion device when performing monitoring presence on the power line voltage and insulation resistance to ground power line; AD - affixion device designed to communicate with the device disconnected line recloser; IMU - insulation monitoring unit that provides measurement and evaluation of the insulation level to ground disabled power line; CUCB - the control unit circuit breaker of served affixion /2/.

The developed automatic reclosing devise consists of three main components: insulation monitoring and control unit; connection device; the sensor of the high voltage existence, intended for installation in cells of mobile and stationary quarry distributive substations of power supply systems with the insulated and compensated neutral and with a working voltage of 6...10 kV. The device can come into

action from operation of protection against short circuits on the ground and, if necessary, the maximum current protection.

To predict the level of electrical safety and reliability of electrical grids and equipment in the operation of power systems (any enterprise and the more so quarres) for a number of reasons, primarily one should know the condition of their insulation. Continuous and automatic values control of insulation components of an electric grid (resistance and capacitance of the network phase insulation to the ground, the inductance of the compensating device) will allow predicting emergence of dangerous system conditions and, at available technical capability, to enter the advancing operating commands, allowing to minimize possible damage. For the specified purposes essentially new way of continuous measurement of component values of insulation resistance related to the ground of an electric grid and its elements under the working voltage is offered.

The functional diagram of continuous measurement system of related to the ground insulation parameters of all distributive grid or its elements with-   out removal of working voltage is shown in Fig. 2.

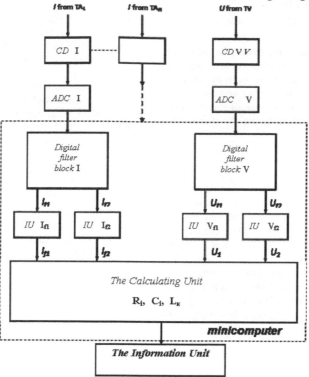

Figure 2.The functional diagram of continuous measurement system of related to the ground insulation parameters

The system structure assumes a voltage channel and a few (controlled by the affixion number) of current channels. Voltage channel, on which entrance the signal arrives from the TV voltage (voltage measurement transformers), connected to the busbars of the switchgear, is designed for removing, handling and measuring the values of operating voltages U_{f1} and U_{f2} and transfer them to the computer module for future use. Current channels, on which input signals are received from the TA$_i$ current sensors (current measurement transformers) for the outgoing feeders, are also designed to remove, process and measurement values and operating currents I_{f1} and I_{f2} of the distribution grid respective sections and transfer them to the computer module.

These channels are differ by entrance blocks (coordinating devices) CD providing linking of analog-digital converters and the corresponding measuring sensors (transformers). When the signal processing system performs the following operations in sequence: the analog signals are converted to digital (ADC); the corresponding blocks of a digital filtration (are intended for selection by program methods from the general current signal of operational frequencies signals) allocated and sepa-

rated of operating frequency components; signals are measured (IU - quantification of the signals corresponding frequencies) and their values are entered in the calculating unit of the system.

The calculating unit on the basis of both measured and recorded values of operating voltages and currents from the expressions (13) with the transformation coefficients voltage measuring transformers k_V and current transformers k_{Ai} at given points of the distribution grid of power supply system calculates:

- insulation resistance to ground of the three phases corresponding to the entire grid or controlled section (i-th affixion)

$$R_i = \frac{U_1 U_2}{k_V k_{Ai}} \sqrt{\frac{\omega_2^2 - \omega_1^2}{U_2^2 I_{1i}^2 \omega_2^2 - U_1^2 I_{2i}^2 \omega_1^2}} \ ;$$

- the total capacity of all three phases of the grid or the corresponding controlled section (i-th affixion) related to the ground

64

$$C_i = \frac{k_V k_{Ai}}{U_1 U_2} \sqrt{\frac{U_2^2 I_{1i}^2 - U_1^2 I_{2i}^2}{\left(\omega_1^2 - \omega_2^2\right)}} \ .$$

The information unit of the system is intended to the calculation results representation in a form convenient for specific conditions or form message and sending it to the power dispatcher and other interested services. In general, the use of microcontrollers or the microcomputers allows the offered method to use:

- for an operational measurement of insulation resistance level of both all power network as a whole, and each of affixion of a distributive grid;

- for an operational measurement of related to the ground capacity level of both all power network as a whole, and each of affixion of a distributive grid;

- for an operational measurement of arc suppression coil inductance values (compensating device);

- for automatic compensating device adjust in a resonance with the distributive grid capacity;

- to carry out selective leakage protection, protection against earth fault or alarm in power systems (quarries and mines), regardless of the configuration and mode of neutral network.

CONCLUSIONS

1. The new principle of protection method work is given, which is based on the performance evaluation of zero sequence current vectors and phase voltage vectors relative position, which can significantly increase the sensitivity of protection devices against ground faults and make it virtually independent of the total network capacity relative to the ground.

2. The theoretical substantiation is given and the method of ground insulation parameters selective determination of an electric network under the operating voltage is developed, based on imposing on a distributive grid of bi-frequency operational sinusoidal signals of noncommercial frequency.

3. The function diagram of continuous and automatic control system or relative to the ground insulation resistance and capacity measurement for a three-phase electric network over 1000 V is presented.

4. Depending on a place of current measuring sensors turning on the system is able to provide selective measurement of the whole power grid ground insulation or the corresponding affixion, also the arc suppression coil inductance.

REFERENCES

Pivnyak G.G., Sckrabets F.P. 1993. *Asymmetrical damage in electrical networks Career: A Reference Guide* (in Russian). Moscow: Nedra: 192.

Pivnyak G.G., Sckrabets F.P., Gorbunov Ya.S. 1992. *Relay protection of electrical installations in open cast mining: A Reference Guide* (in Russian). Moscow: Nedra: 240.

Pivnyak G.G., Sckrabets F.P., Zayika V.T., Razumniy Yu.T. 2004. *Systems of energy efficient coal mines (monograph)* (in Ukrainian). Dnipropetrovs'k: National Mining University: 203.

Tompkins W., Webster D. 1992. *Interfacing sensors and input devices to computers IBM PC* (transl. from English). Moscow: Mir: 592.

Gamkrelidze S.A., Zavjalov A.V., Malstev P.P. & Sokolov V.G. 1988. *Digital processing of information on the basis of high-speed LSI* (in Russian). Moscow: Energoatomizdat: 136.

Power Engineering, Control and Information Technologies in Geotechnical Systems – Pivnyak, Beshta & Alekseyev (eds)
© 2015 Taylor & Francis Group, London, ISBN 978-1-138-02804-3

Comparative analysis of methods for estimating the Hurst acoustic signal whenever feed rate control in jet mills provided

M. Alekseyev & L. Berdnik
State Higher Educational Institution "National Mining University", Dnipropetrovs'k, Ukraine

ABSTRACT: The Hurst exponent has been estimated by the methods of scaled range and dispersion change in the aggregated series of the acoustic signal in a jet mill taking into account the variety of feed rate offsets.

1 INTRODUCTION

High demands on the quality of the materials which often contain powders that improve strength, hardness and durability of these materials, is increasingly the subject of special attention in modern industry.

To maintain fine and superfine grinding, jet method of comminution is widely used in industry. The material grounded in a jet mill can have new properties. The necessity to increase the productivity of mills aimed at obtaining high-quality fine powders requires more precise feed rate control in the mill.

2 ANALYSIS OF THE ACHIEVEMENTS AND PUBLICATIONS IN THIS FIELD

To determine the rate value of in-mill feeding, the methods based on the analysis of the acoustic signal of the mill are widespread nowadays. In (Alekseyev, Holod, Timchenko 2012),the Hurst acoustic signal assures feed rate control in a jet mill.

Given experimental data, the Hurst exponent estimation plays an important role in studying the processes with properties of self-similarity. All methods of the Hurst exponent estimation still have certain disadvantages. In particular, all the estimates of the indicator are biased random variables.

To date, a number of methods for estimating the Hurst exponent have been developed on the basis of empirical data. These methods prove different accuracy of the results obtained.

3 THE OBJECTIVE AND TASKS OF THE RESEARCH

The aim is to study the correlation between an in-mill feed rate of a jet mill and the value of the Hearst exponent of the acoustic signal obtained by the method of scaled range (R/S-analysis) and dispersion change in the aggregated series.

4 STATEMENT OF THE RESEARCH BASICS

In this paper, two computing methods as for the Hurst exponent are considered.

The scaled range method involves the following:

1. Transform non-stationary time series of length M into the time series of length N = M-1;

2. Divide the time interval N by adjacent subintervals A of length n, so that $A*n = N$. Where in N, A and n are chosen in such a way that $\frac{M-1}{n}$ is an integer value. Mark each subperiod I_a given that $a = \overline{1,A}$. Introduce the dual code, i.e. each element of N_i in I_a is given a symbol $N_{k,a}$, $k = \overline{1,n}$. For each I_a of length n, the average value over the interval is determined by:

$$e_a = (^1/_n) \sum_{k=1}^{n} N_{k,}\,;$$

3. Calculate the elements of the time series for accumulated deviations $(X_{k,a})$ from the mean value for each subinterval I_a:

$$X_{k,a} = \sum_{i=1}^{k} \left(N_{i,a} - e_a \right), k = \overline{1,n}$$

4. The range is determined as the maximum value minus the minimum value of $X_{k,a}$ within each subinterval I_a:

$$R_{I_a} = \max(X_{k,a}) - \min(X_{k,a}), 1 \le k \le n$$

5. Sample standard deviation is calculated for each subinterval I_a:

$$S_{I_a} \sqrt{(\tfrac{1}{n}) \sum_{k=1}^{n} (Nk,a - ea)^2};$$

6. Each range R_{I_a} now is normalized by dividing it by the corresponding S_{I_a}. In step (2), we get adjacent subperiods of length n. Hence, the mean value *(R / S) n* is determined by:

$$(R/S)_n = (1/A) \sum_{a=1}^{A} \left(\frac{R_{I_a}}{S_{I_a}}\right);$$

7. Length n iteratively increases up to the next greater value. Steps (2-6) are repeated until $n = (M-1)/2$. Obtain two vectors $\bar{n}$ and $\overline{R/S}$.

8. Then, on the basis of the values obtained in step (7),apply the regression in the equation $(R/S)_n = c*n^H$, where $c = const$, and H–the Hurst exponent. Find the logarithms for both sides of the regression equation beforehand. The intercept on a coordinate axis is an estimate of log *(c)*, a constant. The line slope is an estimate of the Hurst exponent H.

To employ the method of dispersion change in the aggregated series, the following steps are undertaken.

The aggregation on the time scale with the parameter m is considered as a transition to the process *x (m)*, that

$$x_k^{(m)} = \frac{1}{m} \sum_{i-km-m+1}^{km} x_i$$

For aggregated time series *x (m)*of self-similar process, dispersion at great values of m is calculated by the formula:

$$Var\left(x^{(m)}\right) \sim \frac{Var(x)}{m^\beta}$$

In this case, the self-similarity parameter H = (1-β)/2 can be determined if we succeed both in generating the aggregated process at different aggregation levels m and in calculating the dispersion for each level.

The plot of $log(Var(x^{(m)}))$ versus *log (m)*is the straight line with the slope equal to -β.

If the value of $H = 0.5$, the time series is called "white noise." The values of the series are random and uncorrelated. The future values do not depend on the present ones. The probability density function can be the normal curve, but this is not a prerequisite. The R/S-analysis classifies an arbitrary series, irrespective of what kind of distribution it corresponds to.

The values of the Hurst exponent in the range of 0<H <0.5 characterize the antipersistent time series, which is called "pink noise". These processes are most typical for the effects of turbulence. This type of time series often is called "reversion to the mean." If the series shows a "growth" in the previous period, it is likely to decrease in the next period. Conversely, if it went on to decrease, it is close to the rise. The sustainability of such antipersistent behavior depends on how much N is close to zero. Such a series is more volatile than the random one, as it consists of rapid drop-rise reverses.

The values of the Hurst exponent in the range of 0.5<H <1 characterize the persistent time series. Such processes are referred to as "black noise". The persistent time series shows inherent trend stability. If the series increases (decreases) in the previous period, it is likely that this trend will maintain for some time in the future. The closer to 0.5 N, the noisier the series is, and the less explicit the trend is. The persistent series is a generalized Brownian motion, or biased random walks. The strength of this bias depends on whether or not N is greater than 0.5.

There is also a fourth characteristic of the Hurst exponent, where N>1. In this case, one considers the Levi statistics and the process (or time series) with fractal time, the time points of discontinuity of the derivative. This means that the amplitude jumps occur independently and take place, according to Levi, during the time specified by the value of the jump, and grow with it. The variance of the increment for the given time interval becomes final, the trajectory in the phase space preserves its form, but a new fractal object appears, i.e. time points of discontinuity of the derivative.

To assess the potentialities inherent in the Hurst exponent of acoustic signals obtained by different methods, acoustic signals at various stages of feed control in a jet mill have been analyzed: the empty mill, the mill being charged, the operation mode, the mill being discharged.

The resultsof characterization the Hurst acoustic signal in a jet mill by the method of scaled range are given below.

Fig. 7 and Fig. 8 show both the signals of the acoustic signal in the mill at different modes of operation and the approximating lines for characterization the Hurst parameter.

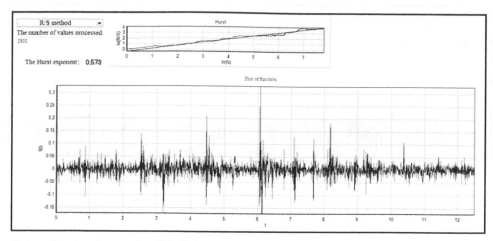

Figure1. Acoustic noise whenever the mill is discharged

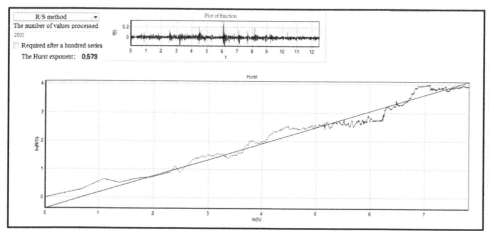

Figure 2. Approximating line to characterize the Hurst exponent as for the mill being discharged

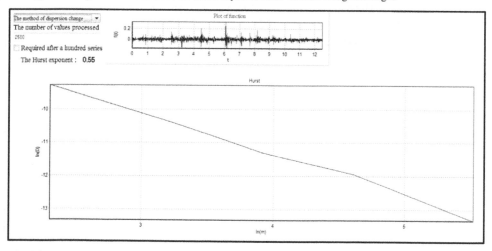

Figure 3.Characterization of the Hurst exponent by the method of dispersion change as for the jet mill being discharged

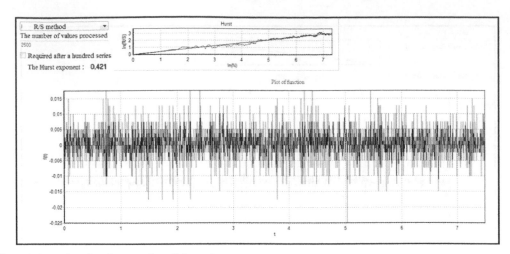

Figure 4. Acoustic noise whenever the mill is empty

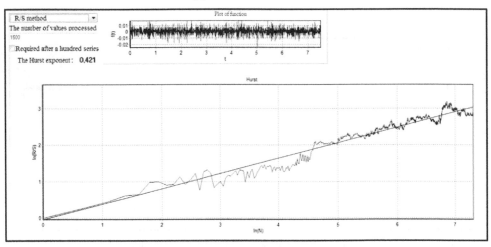

Figure 5. Approximating line to characterize the Hurst exponent as for the empty mill

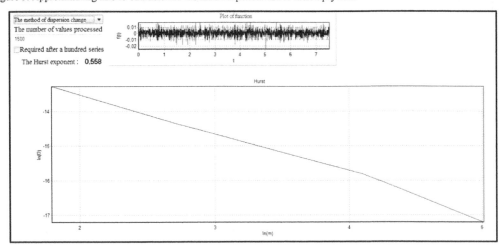

Figure 6. Characterization of the Hurst exponent by the method of dispersion change as for the empty jet mill

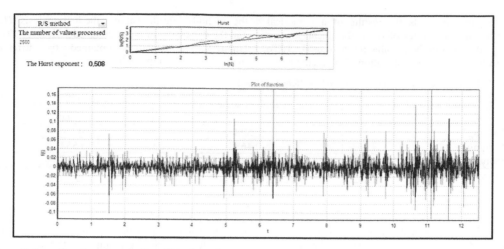

Figure 7. Acoustic noise whenever the mill is charged

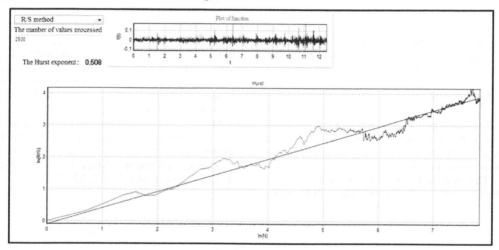

Figure 8. Approximating line to characterize the Hurst exponent as for the mill being charged

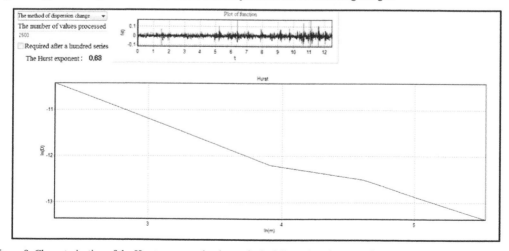

Figure 9. Characterization of the Hurst exponent by the method of dispersion change as for the jet mill being charged

The paper presents the results of a numerical experiment. The values of H for the lineup tend to change throughout the possible range of $0<H<1$. To obtain the time series, estimation H is calculated by the methods described above: the R/S-analysis (H_{rs}) and the dispersion of the aggregated series (H_d). The Hurst exponent values obtained by different methods are given in Table 1.

Table 1. The Hurst exponent values obtained by different methods

Operation mode of the mill	The method of scaled range	The method of dispersion changes
Charging1	0,508	0,63
Charging2	0,468	0,554
Empty1	0,421	0,558
Empty2	0,458	0,57
Discharging1	0,573	0,55
Discharging2	0,481	0,614

The Hurst exponent values (obtained by the method of scaled range) for the analyzed modes are: before the charging $H = 0,421$; charging $H = 0,508$; operating mode $H = 0,573$; discharging $H = 0,481$. The results obtained within the real time series prove that the Hurst index is essential for classification of the acoustic signal to determine the in-mill feeding rate as for a jet mill.

5 CONCLUSIONS

1. The study of such characteristics as the Hurst exponent helps predict dynamics of the acoustic signals, which reflect the processes of grinding of the material in a jet mill.

2. The proposed approach for monitoring and controlling the rate of charging a jet mill by means of the Hurst exponent of the acoustic signal contributes to the grinding performance and grindability index.

REFERENCES

Alekseyev M.O., Holod E.L., Timchenko E.M. 2012. *Estimation of the in-mill feed rate in jet mills on the basis of the Hurst exponent of the acoustic signal* (in Russian). Scientific works collection of NMU, Issue 37: 179-184.

Butakov B., Grakovsky A. 2005. *Estimation of stochasticity levels for time series of arbitrary origin by means of the Hurst exponent* (in Russian). Computer Modeling and New Technologies, Issue 9, Vol. 2: 27-32.

Automatic control of coal shearer providing effective use of installed power

V. Tkachov, N. Stadnik & A. Bublikov
State Higher Educational Institution "National Mining University", Dnipropetrovs'k, Ukraine

ABSTRACT: Idea of coal shearer automated control is substantiated in terms of coal cutting and loading with the help of auger using a criterion of power rescheduling. Ratios of cutting power and loading power at different coal shearer velocities as well as coal cutting resistance have been determined. Dependence of maximum velocity of a coal shearer as for sustainable electric drive power on auger rotational frequency is demonstrated. A technique to divide power by cutting drive into components connected with a process of coal cutting and loading with the help of auger is proposed. An approach to determining commanded and feeding velocity levels on the criterion of coal cutting and loading power rescheduling with the help of auger has been substantiated.

1 INTRODUCTION

Nowadays almost each automatic control system for coal shearer implements minimum automation alternative when two operation modes with automatic transition from one to another is naturally supported (Kowal, Podsiadło, Pluta & Sapiński 2003):

- stabilization of power level for cutting drive electric motor at the expense of changes in feeding velocity involving conveyer transport capabilities limitations, permissible methane concentration, and other factors;
- stabilization of feeding velocity in terms of constant cutting speed involving limitations on electric motor drive overloading capability as well as above limitations.

It should be noted that the algorithm of automated control has been developed for coal shearers having high loading capacity of their end organs. It is also effective for coal shearers operated within middle and high seams. However the algorithm cannot involve advantages of coal shearer cutting velocity control to be used for its efficiency increase as well as for decrease in specific power consumption for cutting, transportation, and coal loading with the help of end organ.

2 STATEMENT OF THE PROBLEM

To mine coal within seams which thickness is up to 1 m coal shearers with auger end organs are mainly used. Old-generation coal shearers capable to have displacement velocity up to 5-6m/min

consume up to 30% of total capacity of cutting electric motor drive (Boiko 2002). Modern coal shearers are characterized by high power supply capacity; at higher displacement velocities their auger needs increased power to load coal resulting in decrease of cutting power. The fact limits coal shearer efficiency. Due to it a problem of rational use of modern coal shearer high power supply capacity arises when it is required to provide maximum efficiency owing to automated control algorithm improvement.

Since operating conditions of coal shearer are in a constant change, a system of a coal shearer automated control should be adaptable setting the displacement velocity when coal shearer operates with complete implementation of its power supply capacity and optimum economic parameters.

The paper aims at substantiating of automated control algorithm for a higher power supply capacity coal shearer to improve its efficiency using rescheduling of components of cutting motor drive power being spent for coal cutting on the one hand and its transportation and auger loading on the other one.

3 THE RESEARCH TOOL

The research is carried out using "face – auger – cutting electric power drive" simulation model which basis is as follows: techniques of power characteristics calculations for coal shearer (Pozin, Melamed & Ton 1984), mathematical description of energy conversion in electric drive (Starikov, Azarkh & Rabinovich 1981), results of numerous studies of coal shearer statistic dynamics to simulate end organ loads (Dokukin, Krasnikov & Khurgin 1978), and calculation procedure of power characteristics for auger coal transportation and loading (Boiko 2002). The simulation model is described in the paper (Tkachov, Bublikov & Isakova 2013).

4 THE RESEARCH RESULTS

As is known, total power of coal shearer cutting drive is consumed for coal crushing and transportation by auger. Define the values of these powers depending on a feeding velocity. Moreover, show the way of changing total power at alternating auger rotating frequency for following conditions: seam thickness – 0.85 m, auger diameter – 0.8 m, bedding angle – 35°, coal is tough (see Fig.1).

Dependence of cutting power on feeding velocity at different coal cutting resistance values (Pozin, Melamed & Ton 1984) is as follows:

$$P_p = \frac{F_u V_p}{1000 \eta_p}, \qquad (1)$$

where P_p is cutting power, kW; F_u is total auger cutting power , N; V_p is cutting velocity, m/s; η_p is efficiency of cutting drive reducing gear.

Cutting drive motor power consumed for coal loading is determined by (Pozin, Melamed & Ton 1984):

$$P_{\text{norp}} = \frac{\pi F_{\text{norp}} D_u n_{\text{o6}}}{60 \cdot 1000}, \qquad (2)$$

where P_{norp} is cutting power, kW; F_{norp} is power of material loading resistance, N; D_u is auger diameter, m; n_{o6} is auger rotational velocity, rot/min.

Fig.1 demonstrates dependences based on (1) and (2).

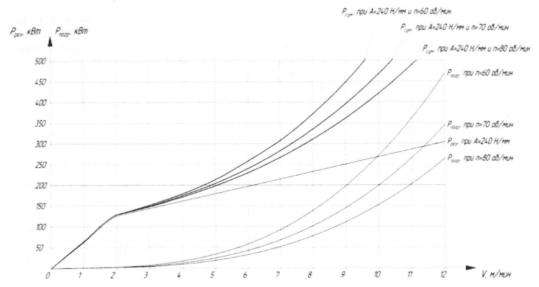

Figure 1. Dependences of cutting and loading power on feeding velocity if coal cutting resistance is 240 N/mm

Cross of sustainable motor power lines and cutting power lines (Fig.1) allows determining maximum feeding velocity on the factor of motor overloading capacityfor a cutting component without component of power consumed for loading. If coal cutting resistance is 240 N/mm (Fig.1) thenfeeding velocity of a coal shearer on the givenpower component is limited at the level of 6 m/min.

Crosses of power loading curves with the line of sustainable power of a motor cutting drive determine maximum feeding velocity of a coal shearer on the factor of a motor overloading capacity for a loading component without taking into account the power consumed for cutting.

Consider effect of auger rotational velocity on loading power. Calculated loading curves explain that when rotational frequency increases, a curve shifts rightwards. That increases maximum displacement velocity on the factor of a motor overloading capacity for loading component (Fig.1). ON the factor maximum velocity if 9, 10, and 11 m/min if auger rotational frequency is 60, 70, and 80 rot/min respectively.

74

A sum of (1) and (2) expressions gives a formula to calculate total power of a cutting drive motor:

$$P = \frac{\pi n_{o6} D_u (F_u + F_{norp} \cdot \eta_p)}{60 \cdot 1000 \cdot \eta_p}. \qquad (3)$$

Total power of cutting drive motor should not be more than its sustainable power:

$$P_y = \frac{M_y \cdot n_y}{9550}, \qquad (4)$$

where M_y is sustainable torque for cutting drive, N·m; n_y is rotational frequency of a motor adequate to a sustainable torque, rot/min.

Crosses of total powers with a line of sustainable power determine maximum velocities of coal shearer displacements on overloading capacity of a motor. Make a table (Table 1) of these values for different frequencies of auger rotations and coal cutting resistance (A). Graphs in Fig.2 show that when coal cutting resistance is 360 N/mm then loading power cannot be a limiting factor for a coal shearer as its velocity is not more than 2 m/min; such loading velocities consume minor share of a motor cutting drive total power.

Table 1. Dependence of a coal shearer displacement velocity (m/min) on an auger rotational frequency and coal cutting resistance

n, rot/min	A, N/mm		
	120	240	360
60	6.4	4.6	1.75
70	8	4.8	1.75
80	9.6	5	1.75

Coal cutting resistance of the majority of coal reserves is up to 240 kN/m (almost 89%). Hence in most cases increase in coal shearer displacement velocity is effective owing to increase of auger rotational frequency.

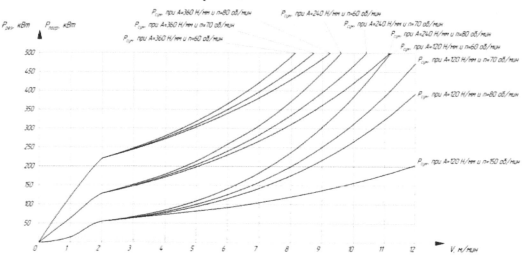

Figure 2. Dependence of total cutting drive power on feeding velocity in terms of different values of coal cutting resistance and auger rotational frequency

To increase coal shearer displacement velocity one should increase an auger rotational frequency by certain value; to determine it recalculation of loading power is as follows:

$$P = P_{HOM} - P_V, \qquad (5)$$

where P_{HOM} is cutting drive nominal power; P_V is cutting power in terms of accelerated velocity of a coal shearer displacement, kW.

Using obtained value of loading power (2), determine value of desired rotational frequency:

$$n_{o6} = \frac{60 \cdot 1000 \cdot P_{norp}}{\pi F_{norp} D_u}. \qquad (6)$$

Analysis of graphs (Fig. 1 and Fig.2) shows that coal shearer displacement velocity is limited by displacement drive power, cutting power, and loading power. When displacement velocity increases, loading power velocity increases as well; as a result at 6 to 7 m/min velocities it is more than 30% of cutting drive total power. Increasing an auger rotational frequency may factor into its decrease while mining seams with up to 240 N/mm coal cutting resistance at available powers of cutting drives.

When an auger rotational velocity changes feeding velocity is controlled as follows. Suppose that total consumed power increase along with a coal shearer displacement (Fig.3). In this context a curve of total power which was initially similar to a curve 1 takes a shape of curve 2. To continue coal shearer operation at V_1 velocity, power increasing the initial one by ΔP_{CYM} value is required.

$$\Delta P_{CYM} = \frac{\pi n_{o6} D_{\text{и}}(F_{u1} + F_{\text{погр}}\eta_p)}{60 \cdot 1000 \eta_p} - \frac{\pi n_{o6} D_{\text{и}}(F_{u2} + F_{\text{погр}}\eta_p)}{60 \cdot 1000 \eta_p} = \frac{\pi n_{o6} D_{\text{и}}}{60 \cdot 1000 \eta_p} \cdot (F_{u1} - F_{u2}). \qquad (7)$$

Taking into consideration the fact that nominal values limit total power value, feeding velocity of a coal shearer should be reduced. Velocity decrease $V_1 - V_2 = \Delta V_c$ (Fig.3) results in proportional decrease in efficiency:

$$\Delta Q_C = Q_1 - Q_2 = (V_1 - V_2) \cdot B_3 \cdot H \cdot \gamma_{OB}. \qquad (8)$$

Consumed power value can be reduced at the expense of increase in an auger rotational frequency. It allows increasing coal shearer displacement velocity by $V_3 - V_2 = \Delta V_{\text{п}}$ value (Fig.3); in turn it provides efficiency increase:

$$\Delta Q_\Pi = Q_3 - Q_2 = (V_3 - V_2) \cdot B_3 \cdot H \cdot \gamma_{OB}. \qquad (9)$$

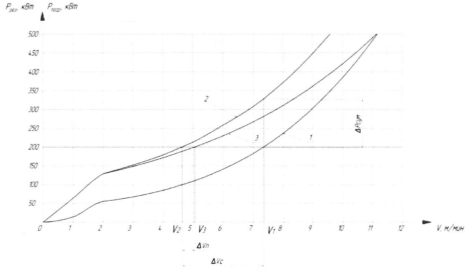

Figure 3. Determining commanded velocity level for coal shearer displacement velocities as well as an auger rotation basing upon operating statistic specifications of a coal shearer

Available algorithms based upon total load value control (without separating components of power consumed by cutting drive electric motor) can only provide coal shearer displacement velocities. Increase in total load decrease feeding velocity and efficiency respectively.

The proposed algorithm (Fig.3) makes it possible to obtain increased efficiency value (9) to compare with available algorithms.

An auger rotational frequency control is possible owing to application of frequency-regulated converter. Today frequency converter of a coal shearer drive displacement for thin seams is placed

within entry. It is expedient to refine its design by building in frequency converter for a cutting drive. Moreover increased velocity of an auger rotation (for example, 150 rot/min) at the expense of adequate power ratio of reducer should meet nominal frequency of 50 Hz supply voltage. Then constant sustainable torque of cutting drive motor will be observed while controlling an auger rotational velocity within the whole range of its variation.

Calculation algorithm for commanded feeding and cutting velocities is an important point for implementation of the approach aimed at coal shearer efficiency increase. Namely it is important to calculate values to be changed for to achieve required changes in cutting and loading powers. Traditional method of load control according to the available value of cutting drive motor power cannot be applied as a criterion due to the availability of two components as it has been shown above.

Here is the method to determine commended feeding velocities and auger rotational velocities using a system of automated control to change power of cutting drive electromotor power by the given value on the basis of incrementation of cutting and loading powers forecasting. This method is proposed as an integral part of approach development for mining processes classification using information and intellectual means being the basis for implementation of one of the function of intellectual support of making decisions while coal shearer controlling. The information on changes in cutting drive electromotor power connected with various physical processes (rock mass cutting with the help of picks and loading broken mass by auger blade) will help to imagine nature of the physical processes. Besides it will help to have more efficient control of coal shearer from the viewpoint of efficiency and power consumption.

Regularities being the basis of the technique have been obtained using "face – auger – cutting electric power drive" simulation model (Tkachov, Bublikov & Isakova 2013) for УКД300 coal shearer.

It is proposed to single out components of cutting drive electromotor power basing on defining increment coefficients of average cutting and loading powers per unit of coal chippings thickness (in the middle of crescent-shaped section) while coal shearer operating:

$$K_C = \frac{\Delta P_C}{\Delta h} \; ; K_L = \frac{\Delta P_L}{\Delta h}, \, kW/cm, \qquad (1)$$

where ΔP_C and ΔP_L are values of changes in cutting drive electromotor power (CDEP) consumed for coal crushing (cutting power) and coal transportation and loading using auger (loading power) when chipping thickness varies by Δh, kW.

Determination of increment coefficient of only one power component needs changes in the given component (when chipping thickness varies); another component stays invariable. Consider two operation modes of coal shearers meeting the requirement and develop adequate static characteristics.

Fig. 4 explains that when velocities of cutting and feeding are of correlated changes (to provide constant chipping thickness) CDEP power varies in a feeding velocity function practically according to the same law as cutting power does. That depends on the fact that nature of coal transportation and loading using auger at cutting and feeding velocities increase experience minor variations as similar coal amount should be fed into working volume of a auger per one its rotation. That is why power consumed for coal transportation and loading using auger experience minor variations to be confirmed by Fig. 4,b.

Since correlated changes in velocities of auger rotation (N) and feeding (V) depending upon CDPE feeding power velocity and cutting power are the same then CDPE power increment can be used to determine changes in cutting power:

$$\frac{\Delta P_C}{\Delta V_{(V/N-const)}} \approx \frac{\Delta P_M}{\Delta V_{(V/N-const)}}. \qquad (2)$$

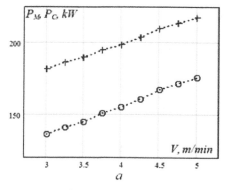
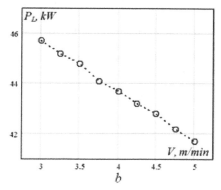

Figure 4. Static dependence on feeding velocity (V) in terms of constant chipping thickness (2.5 cm): a is active power of electromotor cutting drive (P_M are markers in the form of crosses) and cutting power (markers in the form of circles); b is loading power

Fig.5 demonstrates that increment of CDPE power and cutting power are practically similar until feeding velocity is incremented by 0.75 m/min. In this context relative deviation of cutting power increment from CDPE power increment is not more than 10%. Further difference between increments of cutting and CDPE power starts increasing due to loading power effect (Fig.5). Due to the effect of high-frequency component of CDPE which variation coefficient is 0.12 power averaging within relatively short time period (up to 15 sec) takes place with a random error which maximum values

are ± 1 kW. As a result CDPE power increment coefficient in terms of feeding and cutting velocities increase is better to determine for maximum increment of feeding velocity. However as it is mentioned above if feeding velocity increment is more than 0.75 m/min, unallowable relative error of more than 10% arises due to loading power effect in determining cutting power increment on CDPE power. Hence it is recommended to use 0.75 m/min feeding velocity increment to determine cutting power increment coefficient through CDPE power if feeding and cutting velocities increase.

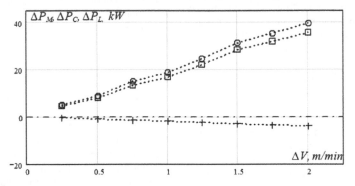

Figure 5. Static dependences of active power increment of cutting drive electromotor (markers in the form of squares) as well as cutting power (markers in the form of circles) and loading ones (markers in the form of crosses) upon feeding velocity increment in the mode of constant chipping thickness control

Note that despite the constant chipping thickness increase in feeding and cutting velocities results in cutting power increase according to linear law owing to increase in auger pick path length. However, if we increase only auger rotational velocity and nature of coal crushing with picks experiences minor variations (which takes places for available augers if only chipping thickness is more than 2 cm) then cutting power stays practically constant as increase in auger pick path length compensate decrease in chipping thickness. Consequently, variations in cutting power in terms of changes in auger pick path length (in the context correlated control of feeding and cutting velocities) help forecasting its changes under corresponding variations of chipping thickness when we pass from one auger rotational velocity to another one:

$$\frac{\Delta P_C}{\Delta N_{(V/N-const)}} \approx \frac{(\Delta P_C)_h}{\Delta N_{(V-const)}}.$$

It means that we can forecast change in cutting power when feeding velocity either increases or decreases having constant velocity of auger rotation if only chipping thickness experience similar change:

$$\frac{(\Delta P_C)_h}{\Delta N_{(V-const)}} = \frac{(\Delta P_C)_h}{\Delta V_{(N-const)}}.$$

For example, if feeding velocity increases from 4 to 4.75 m/min and auger rotational velocity similarly increases from 80 to 95 rot/min cutting power increased by 16.5 kW to keep chipping thickness at the level of 2.5 cm (Fig.4,a). If auger rotational velocity increases from 80 to 95 rot/min then chipping thickness decreases by:

$$\Delta h = 0.278 \cdot 180 \cdot (\frac{4}{80} - \frac{4}{95}) = 0.395 \, cm.$$

Hence chipping thickness decreases by the same value if in terms of auger rotation velocity of 95 rot/min feeding velocity is decreased from 4.75 m/min to 4 m/min:

$$\Delta h = 0.278 \cdot 180 \cdot (\frac{4.75-4}{95}) = 0.395 \, cm.$$

Consequently, if feeding velocity decreases from 4.75 m/min to 4 m/min and auger rotational velocity is 95 rot/min then cutting power is decreased by 16.5 kW.

78

Thus, CDPE power increment in the context of correlated control of feeding and cutting velocities helps forecasting change in cutting power if only feeding velocity is controlled:

$$\frac{\Delta P_C}{\Delta V_{(V/N-const)}} \approx \frac{(\Delta P_C)_h}{\Delta V_{(N-const)}}. \tag{3}$$

It should be noted that cutting power increment coefficient (K_C) is sensitive to coal cutting resistance (A), as cutting power depends proportionally on the parameter. Hence if some period involving important changes in coal hardness has passed then the coefficient should be adjusted. The result of direct proportional dependence between cutting power, feeding velocity, and coal cutting resistance is a formula helping forecast cutting power increment in terms of previous feeding velocity V_1 if coal cutting resistance changes from A_1 to A_2 on the current feeding velocity V_2:

$$(\Delta P_C)_A = V_1 \cdot \frac{(P_C)_{A2} - (P_C)_{A1}}{V_2} \approx V_1 \cdot \frac{(P_M)_{A2} - (P_M)_{A1}}{V_2}. \tag{4}$$

Owing to the fact that loading power stays practically invariable in terms of coal hardness changes (to be shown below) adjustment of cutting power increment coefficient at changes in coal cutting resistance may be performed on CDPE power. Then cutting power increment coefficient for new coal cutting resistance (A_2) may be calculated by:

$$(K_C)_{A2} \approx \frac{(\Delta P_C)_V + ((P_M)_{A2} - (\Delta P_C)_V - (\Delta P_L)_V) - ((P_M)_{A1}^{V1} + (\Delta P_C)_A)}{\Delta h}. \tag{5}$$

where Δh is the value of chipping thickness change while passing from feeding velocity V_1 to feeding velocity V_2 (auger rotational velocity is constant скорость вращения шнека), cm; $(\Delta P_C)_V$ and $(\Delta P_L)_V$ are increments of cutting power and loading power respectively while passing from feeding velocity V_1 to feeding velocity V_2, kW;

$(P_M)_{A2}$ is CDPE power when feeding velocity is V_2 and coal cutting resistance is A_2, kW; $(P_M)_{A1}^{V1}$ is CDPE power at feeding velocity V_1 coal cutting resistance A_1, kW.

Second operation mode of coal shearer allowing determination of increment coefficient for one component of CDPE power per unit of coal chipping thickness increment is a mode with constant feeding velocity and variable cutting velocity. Using "face – auger – cutting electric power drive" simulation model (Tkachov, Bublikov & Isakova 2013) we obtain corresponding static characteristics for the given mode of УКД300 coal shearer.

Fig.6 explains that increase in auger rotational velocity (starting from 70 rot/min) results in loading power decrease according to linear law down to 95 rot/min with its following stabilization at the level of 20 kW. Segment of intensive and practically linear decrease in loading power depends first of all on proportional decrease share of force of coal transportation and its loading with the help of auger in its rotational period when cutting velocity increases. Thus, when feeding velocity is 4 m/min and auger rotational velocity is 70 rot/min, corresponding to 2.86 chipping thickness, force coal transportation and loading are 12.1% of auger rotational period. Increase in auger rotational velocity up to 105 rot/min results in chipping thickness decrease down to 1.99 cm and respectively decrease in coal amount fed into working space of auger per its one rotation. In this context share of force of coal transportation and loading using auger within the period of its rotation is 0% now. Further increase in auger rotational velocity improves its loading capability without changing loading power which stays practically constant value for non-force conditions of coal transportation and loading using auger (Fig.6,a).

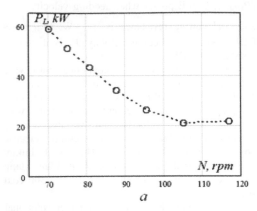
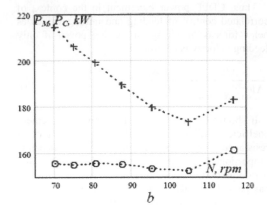

Figure 6. Static dependence of auger rotational velocity when feeding velocity is 4 m/min: *a* is loading power; *b* is active power of electromotor cutting drive (markers in the form of crosses) and cutting power (markers in the form of circles)

As Fig.6,a and Fig.6,b explain a law of CDPE power variation in a function of auger rotational velocity coincides with changes in a loading power law; however, it is true until cutting power experiences serious changes. The condition is met until auger rotational velocity is 105 rot/min; then nature of rock mass crushing with picks changes for the case under consideration. It results in the fact that due to decrease in chipping thickness cutting power experience less intensive decrease and its increase starts prevailing owing to auger pick path length increasing.

Fig.7,a shows possibility of CDPE power use to forecast loading power increment when auger rotational velocity changes:

$$\frac{\Delta P_L}{\Delta N_{(V-const)}} \approx \frac{\Delta P_M}{\Delta N_{(V-const)}}. \tag{6}$$

Fig.7,a explains that up to 17.5 rot/min auger rotational velocity increment, CDPE power increment is practically similar to loading power increment. Further difference arises between increments of the powers due to effect of cutting power which increases along with auger rotational velocity increment increase. Hence to consider accurately CDPE power increment (when relative error is not more than 10%) as loading power increment, in this context auger rotational velocity increment should not be more than 17.5 rot/min.

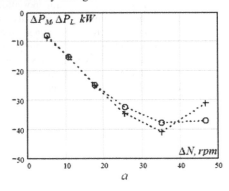
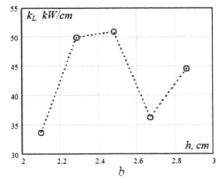

Figure 7. Static dependence of: *a* - increment in loading power (markers in the form of circles) and CDPE active power (markers in the form of crosses) on auger rotational velocity increment if feeding velocity is 4 m/min; *b* - loading power increment coefficient on coal chipping thickness

Fig. 7,b shows dependence of loading power increment coefficient on chipping thickness as a result of CDPE power dependence on auger rotational velocity represented by Fig.6,b (auger rotational velocity range of changes is 70 to 105 rot/min). Despite the fact that in a range of 70 to 105 rot/min for auger rotational velocity CDPE power decreases practically according to linear law (Fig.6,b) and serious divergence in loading power increment coefficient values takes place for different minor segments of the range. It means that static dependence of CDPE power on auger rotational velocity should be obtained with the least stroke of auger rotational velocity which gives possibility to determine rather accurately loading power increment coefficient for any small segment within a range of chipping thickness variation.

Share of force of coal loading and transportation using auger within its rotational period a swell as the character of the processes mainly depends on changes in its working volume, coal amount in it, and opening to unload coal. Thus, since changes in auger working volume and opening to unload coal depend on an auger design, cutting drive body and conveyor side height then the character of coal loading and transportation using auger determines only coal volume fed into auger working space per its rotation, i.e. chipping thickness. Similar changes in chipping thickness may also result from changes in auger rotational velocity at constant feeding velocity and changes in feeding velocity at constant velocity of auger rotation. It means that if dependence of loading power increment on chipping thickness increment is determined using changes in auger rotational velocity at constant feeding velocity then the given dependence may forecast loading power increment while controlling feeding velocity with constant auger rotational velocity:

$$\frac{\Delta P_L}{\Delta h_{(N-var)}} = \frac{\Delta P_L}{\Delta h_{(V-var)}}. \tag{7}$$

Fig.7 demonstrates static dependences obtained at constant feeding velocity being 4 m/min. Analyze validity of the regularities and conclusions as for

similar static dependences determined at other velocities of coal shearer displacement.

Fig. 8,a shows static dependences of loading power on auger rotational velocity obtained for various feeding velocities. Thus, if feeding velocity is 3 m/min we observe practically constant loading power within the whole range of changes in auger rotational velocity. That can be explained by the fact that if feeding velocity is 3 m/min and initial velocity of auger rotation is 70 rot/min, chipping thickness is 2.14 cm when force coal loading and transportation using auger is practically unavailable. As it is sad above further increase in auger rotational velocity cannot change character of coal loading and transportation processes as well as coal loading power.

When feeding velocity is 5 m/min and initial velocity of auger rotation is 70 rot/min, chipping thickness is 3.57 cm and share of force coal loading and transportation using auger is 22.4% within its rotational period. Thus, in this context loading power is higher than it can be seen in Fig.8,a. Note that when auger rotational velocity increases , loading power decreases having the same intensity as for feeding velocity of 4 m/min. That is pass to 5 m/min feeding velocity which increased upper body of the range for chipping thickness change from 2.86 to 3.57 cm has shown the identity of loading power increment for both 1.91 – 2.86 cm chipping thickness change and for 2.86 – 3.57 cm range. It means that there is no significant difference for the considered operation mode of coal shearer which feeding velocity determines loading power dependence (CDPE power) on auger rotational velocity.

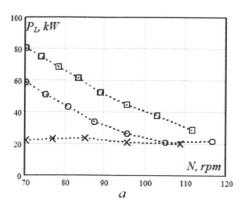

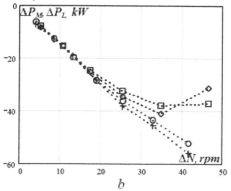

a b

Figure 8. Static dependence of: a – loading power on auger rotational velocity at 3, 4 and 5 m/min feeding velocities (markers in the form of crosses, circles and squares respectively); b – loading power increments (markers in the form of squares for 4 m/min feeding velocities and circles for 5 m/min velocity) and active power of electromotor cutting drive (markers in the form of rhomb for 4 m/min feeding velocities and crosses for 5 m/min velocity) on auger rotational velocity increment

Fig. 8,b confirms conclusions concerning similarity of law of loading power change in the function of auger rotational velocity for 4 and 5 m/min feeding velocities. We can see that before 17.5 rot/min auger rotational velocity increment there are similar increments of loading power increments for both 4 m/min feeding velocity and 5 m/min feeding velocity.

However it should be kept in mind that linear law of loading power variation in a function of auger rotational velocity will take place until changes in coal transportation and loading adequate to redistribution of periods of force and non-force processes of coal transportation and loading within an auger rotational period. Fig. 6.a and Fig.8 explain that linear law of loading power variation in a function of auger rotational velocity is disturbed if chipping thickness is 1.91 cm when share of forces of coal loading and transportation using auger becomes equal to 0% within its rotational period. If this share reaches 100% then similar disturbance of linear law of loading power variation in a function of auger rotational velocity takes place.

Analyze how coal cutting resistance (which average value may differ significantly for various working face sites after coal spall in a seam occurs) depends on loading power dependence on auger rotational velocity. With this purpose the dependence has been obtained for 190 N/mm coal

cutting resistance; Fig.9 demonstrates similar dependence for 140 N/mm coal cutting resistance.

Fig.9 explains that graphs of loading power dependences on auger rotational velocity are practically similar (maximum relative loading power deviation is 3.5% for various coal cutting resistances). Also dependences of loading power increments and CDPE on auger rotational velocity increment are similar for various coal cutting resistances; however it is true only for auger rotational velocities increments up to 17.5 rot/min (Fig. 9,b). There are divergences in these dependences within the areas of auger rotational velocities increment (Fig. 9,b) as there is greater effect of cutting power which in turn greatly depends on coal cutting resistance. Hence we may conclude that dependence of loading power increment on auger rotational velocity increment formed in the process of "training cycle" for certain coal cutting resistance is applicable for any sites of working face with different average coal cutting resistance.

Note that "training cycle" is recommended at the initial stage of seam mining when it is still undisturbed and coal spall does not occur as CDPE power used to determine loading power increment is sensitive to changes in coal hardness along a seam; that introduces an error into definition of loading power increment.

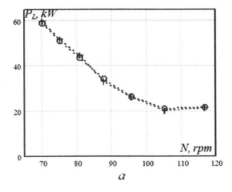
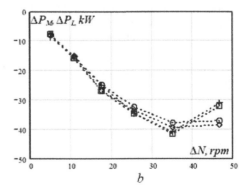

Figure 9. Static dependence of: *a* – loading power on auger rotational velocity (markers in the form of circles for 240 N/mm coal cutting resistance and crosses for 190 N/mm); *b* – loading power increments (markers in the form of circles for 240 N/mm coal cutting resistance and rhombs for 190 N/mm) and CDPE active power (markers in the form of crosses for 240 N/mm coal cutting resistance and squares for 190 N/mm) on auger rotational velocity increment

Below we consider an example how to apply discussed method of feeding velocity settings and auger rotational velocity to vary CDPE power by specified value in terms of automated control of coal shearer in according to following operation modes:

- Redistribution of cutting and loading powers in terms of auger rotational velocity increase;
- Stabilization of CDPE power applying feeding velocity controller and auger rotational velocity controller instead of load controller.

Note that since the method of cutting and feeding velocity settings determination is based on data of average values of CDPE power after each variation of velocity settings they stay unchangeable during a period of CDPE power averaging. Computational experiment based on "face – auger – cutting electric power drive" simulation model has shown that 10 sec is enough to average random high-frequency component of CDPE power with providing 1 kW maximum averaging error. Thus, changes in cutting and feeding velocity settings while controlling coal shearer will be time-discrete with 10 sec period.

Simulation model-based "training cycle" is applied at the initial stage when auger rotational velocity varies within 70-150 rot/min constant 0.1 m increment of chipping thickness (feeding velocity is 4 m/min). Measurements and storage of CDPE average power for each velocity of auger rotation form initial data to determine loading power increment coefficients within various intervals of range of chipping thickness variation. Besides, average CDPE power at 3.5 and 4.25 m/min feeding velocities has been measured within the mode of chipping thickness being at the level of 2.22 cm at the final stage of "training cycle" to determine cutting power increment coefficient while controlling feeding velocity at the level of 4 m/min. It made 171.1 kW and 187.7 kW respectively. Using formulas (1), (2), and (3) calculate cutting power increment coefficient while controlling feeding velocity at the level of 4 m/min when auger rotational velocity is 90 rot/min:

$$\Delta h = 0.278 \cdot 180 \cdot \left(\frac{3.5}{78.75} - \frac{3.5}{95.625} \right) = 0.392 \ cm;$$

$$K_C = \frac{187.7 - 171.1}{0.392} = 42.347, \ kW / cm.$$

Assume that the first mode of coal shearer automated control requires following initial conditions: УКД300 coal shearer feeding velocity is 4 m/min; auger rotational velocity is 80 rot/min; coal cutting resistance is 240 N/mm. Here according to computational experiment using "face – auger – cutting electric power drive" simulation model, CDPE power is 198.6 kW.

Redistribution of cutting and loading powers to reduce loading power requires increase in auger rotational velocity setting from 80 to 90 rot/min when feeding velocity stays unchangeable at the level of 4 m/min. Computational experiment recorded decrease in CDPE power from 198.6 kW to 184.1 kW. Current change in chipping thickness (2.22 – 2.5 cm) based on (1), (6), and (7) formulas needs clarification of loading power increment coefficient:

$$K_L = \frac{\Delta P_L}{\Delta h} = \frac{198.6 - 184.1}{0.278 \cdot 180 \cdot \left(\frac{4}{80} - \frac{4}{90} \right)} = 52.158, \ kW / cm.$$

Feeding velocity setting to increase CDPE average power from 181.1 to 200 kW (CDPE sustainable power) is determined by:

$$200 - 184.1 = \Delta P_L + \Delta P_C = 0.278 \cdot 180 \cdot \frac{V - 4}{90} (K_L + K_C). \quad (8)$$

Equation (8) solution determined setting of a coal shearer feeding velocity: $V = 4.303$ m/. When feeding velocity passes from 4 to 4.303 m/min, coal chipping thickness increases from 2.22 to 2.39 cm. As we can see, in terms of feeding velocity increase chipping thickness variation value differs greatly from similar values when auger rotational velocity varies. That is why loading power increment coefficients matching the given segments of chipping thickness variation range may differ significantly. Hence database formed in the process of "training cycle" is applied to clarify loading power increment coefficient for 2.22-2.39 cm segment of chipping thickness variation range:

$$K_L = \frac{\Delta P_L}{\Delta h} = \frac{191.7 - 184.1}{0.17} = 44.706, \ kW / cm.$$

Further formula (8) taking into account adjusted loading power increment coefficient is used to calculate again setting of coal shearer feeding power being 4.33 m/min. Here value of cutting power increment having been forecasted according to the method is equal to 7.746 kW and loading power is 8.178 kW.

Computational experiment based on "face – auger – cutting electric power drive" simulation model has shown that while passing to 4.33 m/min feeding velocity increase in average CDPE power up to 199.8 kW at the expense of cutting power increase by 7.8 kW and loading power by 7.9 kW. That is CDPE power change error by the specified value is 0.2 kW and relative forecasting errors as for cutting and loading power increments are 0.7 and 3.5% respectively.

According to the latter change in feeding velocity setting taking into account forecasted loading power increment correct cutting power increment coefficient per chipping thickness unit:

$$K_C = \frac{\Delta P_C}{\Delta h} = \frac{(199.8 - 184.1 - 8.178) \cdot 90}{(4.33 - 4) \cdot 0.278 \cdot 180} = 41.104, \ kW / cm.$$

Make one more stage of cutting and loading power redistribution increasing auger rotational velocity from 90 to 100 rot/min. Here 199.8 kW to 185 kW decrease in CDPE average power has been recorded in the process of computational experiment. Loading power increment coefficient is clarified for current changes in chipping thickness (2.17-2.39) based on (1), (6), and (7) formulas:

$$K_L = \frac{\Delta P_L}{\Delta h} = \frac{199.8 - 185}{0.278 \cdot 180 \cdot \left(\frac{4.33}{90} - \frac{4.33}{100}\right)} = 61.4, \ kW / cm.$$

Feeding velocity setting for 185-200 kW increase in CDPE average power involving error of CDPE power variation during previous stage (0.2 kW) is determined by:

$$200 - 185 + 0.2 = \Delta P_L + \Delta P_C = 0.278 \cdot 180 \cdot \frac{V - 4.33}{100}(K_L + K_C). (9)$$

Equation (9) solution determines setting of a coal shearer feeding velocity: $V = 4.62$ m/min. While passing from 4.33 feeding velocity to 4.62 m/min coal chipping thickness experiences $2.17 - 2.315$ cm increase. Loading power increment coefficient for the given segment of chipping thickness range change is clarified by:

$$K_L = \frac{\Delta P_L}{\Delta h} = \frac{191 - 185}{0.149} = 40.27, \ kW / cm.$$

After that formula (9) and adjusted coefficient of loading power increment are used again to calculate setting of coal shearer feeding velocity being 4.703 m/min. Here cutting power increment value forecasted by the method is 7.67 kW and value of loading power is 7.52 kW.

Computational experiment based on "face – auger – cutting electric power drive" simulation model has shown that while passing to 4.703 m/min feeding velocity increase in average CDPE power up to 201.9 kW at the expense of cutting power increase by 8.8 kW and loading power by 8.1 kW. That is CDPE power change error by the specified value is 1.9 kW and relative forecasting errors as for cutting and loading power increments are 12.8 and 7.2% respectively.

According to the latter change in feeding velocity setting taking into account forecasted loading power increment correct cutting power increment coefficient per chipping thickness unit:

$$K_C = \frac{\Delta P_C}{\Delta h} = \frac{(201.9 - 185 - 7.516) \cdot 100}{(4.703 - 4.33) \cdot 0.278 \cdot 180} = 50.18, \ kW / cm.$$

Now consider the case of coal shearer automated control in an operation mode of CDPE power stabilization using feeding velocity controller and auger rotational velocity controller instead of loading controller. Such an operation mode means that when coal hardness increases coal shearer

feeding velocity should decrease to provide equality of actual CDPE average power and sustainable power (200 kW). Cutting and feeding velocities obtained after the second stage of cutting and loading velocities redistributions are initial data for the operation mode. Assume that for certain period, coal shearer has operated with these unchangeable velocities. During the period coal cutting resistance increased from 240 to 270 N/mm resulting in increase of CDPE average power up to 223 kW.

According to (4) and (5) formulas clarify cutting power increment coefficient per chipping thickness unit involving coal cutting resistance variation:

$$(\Delta P_C)_A \approx 4.33 \cdot \frac{223 - 201.9}{4.703} = 19.43, \ kW \ ;$$

$$(K_C)_{A2} \approx \frac{[(223 - 7.52) - (185 + 19.43)] \cdot 100}{(4.703 - 4.33) \cdot 0.278 \cdot 180} = 59.13, \ kW / cm.$$

Feeding velocity setting to decrease CDPE average power from 223 to 200 kW is determined by:

$$223 - 200 = \Delta P_L + \Delta P_C = 0.278 \cdot 180 \cdot \frac{4.703 - V}{100}(K_L + (K_C)_{A2}). (10)$$

Equation (10) solution determines coal shearer feeding velocity setting: $V = 4.199$ m/min. While passing from 4.703 feeding velocity to 4.199 m/min coal chipping thickness experiences $2.35 - 2.1$ cm decrease. Database obtained in the process of "training cycle" helps adjusting loading power increment coefficient for the given segment of chipping thickness variation range:

$$K_L = \frac{\Delta P_L}{\Delta h} = \frac{186.5 - 177.6}{2.35 - 2.1} = 35.18, \ kW / cm.$$

Further formula (10) taking into account adjusted loading power increment coefficient is used to calculate again setting of coal shearer feeding power being 4.216 m/min. Here value of cutting power increment having been forecasted according to the method is equal to 14.41 kW and loading power is 8.57 kW.

Computational experiment based on "face – auger – cutting electric power drive" simulation model has shown the decrease in average CDPE power down to 200.2 kW while passing from 4.703 to 4.216 m/min feeding velocity at the expense of cutting power decrease by 13.6 kW and loading power by 9.2 kW. That is CDPE power change error by the specified value is 0.2 kW and relative forecasting errors as for cutting and loading power increments are 5.9 and 6.8% respectively.

5 CONCLUSIONS

1. The simulation results confirm that within thin seams loading capability of an auger limits a coal shearer displacement velocity. The research explains that 60 to 85 rot/min increase in auger rotational frequency allows increasing maximum velocity of a coal shearer displacement as for an auger loading capability; in turn it makes it possible to improve coal shearer efficiency as well as decrease specific power consumption for coal mining. The paper shows principle capability to implement algorithm of cutting and loading power redistribution in terms of coal shearer automated control on the basis of proposed method of settings for coal shearer feeding velocity and an auger rotational velocity with the help of automated control.

2. Coordinated changes in cutting and feeding velocities to support chipping thickness at constant level, CDPE power increment varies in a function of feeding velocity increment subjecting practically to the same law as cutting power increment if feeding power increment is not more than 0.75 m/min. It can be implemented owing to the fact that the nature of coal transportation and loading using auger stays practically invariable in terms of cutting and feeding velocity increase as similar coal amount is fed into working volume of an auger per one rotation. The regularity can be used to forecast changes in cutting power on cutting drive electromotor power of a coal shearer.

3. When an auger rotational velocity increases, increase in cutting power, due to pick path lengthening, is practically compensated by its decrease owing to coal chipping thickness decrease (it takes place if auger rotational velocity cannot vary a character of rick mass crushing with the help of picks if only chipping thickness is more than 2 cm). Increase in loading power in a function of an auger rotational velocity is increment of cutting drive motor power increment if increment in auger rotational velocity is not more that 17.5 rot/min. The regularity can be applied to forecast changes in loading power on a cutting drive electromotor power.

REFERENCES

Kowal J., Podsiadlo A., Pluta J. & Sapinski B. 2003. *Control systems for multiple tool heads for rock mining*. Acta montanistica slovaca rocnik, Vol.8: 162-167.
Boiko N.G. 2002. *Coal loading by shearers* (in Russian). Donets'k: DonNTU: 157.
Pozin E.Z., Melamed V.Z. & Ton V.V. 1984. *Destruction of coal by stopping machines* (in Russian). Moscow: Nedra: 288.
Starikov B.Ya., Azarkh V.L. & Rabinovich Z.M. 1981. *Asynchronous drive of shearers* (in Russian). Moscow: Nedra: 288.
Dokukin A.V., Krasnikov Ju.D. & Hurgin Z.Ja. 1978. *Statistical dynamics of mining machines* (in Russian). Moscow: Mashinostroenie: 239.
Bublikov A., Tkachov V. & Isakova M. 2013. *Control automation of shearers in term of auger gumming criterion*. CRC Press/Balkema – Taylor & Francis Group: Energy efficiency improvement of geotechnical systems: 137-145.

Power Engineering, Control and Information Technologies in Geotechnical Systems – Pivnyak, Beshta
& Alekseyev (eds)
© 2015 Taylor & Francis Group, London, ISBN 978-1-138-02804-3

Designing algorithm for coal shearer control
to provide decrease in specific power consumption

V. Tkachov, V. Ogeyenko & L. Tokar
State Higher Educational Institution "National Mining University", Dnipropetrovs'k, Ukraine

ABSTRACT: A problem of a shearer cutting drive control is considered. A solution providing specific energy consumption minimization in the process of mining operations is proposed. Difficulties connected with feeding velocity controller development are substantiated. Advanced algorithm scheme to operate a shearer is demonstrated. Research results concerning simulation model to control feeding velocity of cutting drive are analyzed.

1 INTRODUCTION

From the viewpoint of power supply basis for all branches of national economy, mining industry is among the most important in Ukraine. Coal reserves are basic fuel and energy resource. That stipulates high level of being in demand to improve efficiency of powered systems carrying out tasks connected with rock mass separation, transportation, and fragmentation. Implementation of modern sensors and operating mechanisms as well as automation systems helps causing to its rise. Such systems provide increase in efficiency and decrease in specific power consumption for mining processes owing to growth in information capacity of intelligent modules as for the task progress, forecast of events and implementation of improved control algorithms.

2 STATEMENT OF THE PROBLEM

Achieving of maximum coal mining efficiency in terms of predetermined power consumption limit is the main problem while designing automation systems for coal shearers. Its solution is complicated by the necessity of involving a number of factors connected with processes taking place in machine drives and with mined rock mass. No one of available controlled cutting drives can be considered as reliable. Accordingly, stabilization of the main motor drive with the help of feeding device speed variation is the key problem for each operated mining machine (Pivnyak et al. 2007).

Complexity of adequate controller engineering depends on the fact that the number of various factors effect bend point position: hardness of mined material, humidity within humidity area of operation, state of operating device, bluntness of cutting teeth etc. Thus, in the course of coal shearer operation a value of optimum feed velocity ($Vn.k$) experiences constant changes. A graph of consumed power dependence on feeding velocity in terms of various modes explains that (Fig.1).

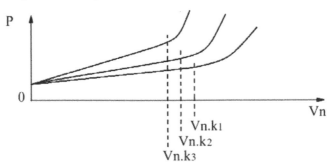

Figure1. Graph of consumed power dependence on feed velocity in terms of various modes

Modern coal shearers control load automatically; however, such an approach can only be used to adhere to given efficiency settings. Power consumption connected with the machine operation is taken into consideration only in connection.

Statement of coal shearer control problem is as follows:

$$\begin{cases} H(Vn) \to \min \\ P \le Po(u) \\ Vn\min \le Vn \le Vn\max \\ Vp = const \\ Vn = var \end{cases},$$

where H(Vn) is specific power consumption, Vn is feed velocity of a coal shearer, P is power consumed by cutting drive, Po(u) is maximum allowable capacity, u is supply voltage, Vnmax/Vnmin is maximum/minimum allowable feed velocity of a coal shearer, Vp is cutting velocity.

Dynamic determination of Vn.k current position basing on leadings of power sensor readings consumed by cutting drive is the key problem of controller operational algorithm.

3 DEVELOPMENT OF ALGORITHM TO CONTROL COAL SHEARER

An approach proposed in paper (Pivnyak et al. 2007) allows determining value for optimum coal shearer feed velocity at current mode. The algorithm is based on step-by-step determination of interval within which linear part of function P(Vn) transforms into non-linear one. Velocity increases/decreases by ΔV value until value of derived function varies (Fig. 2).

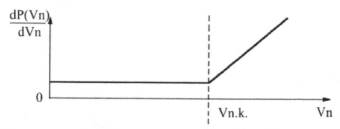

Figure 2. Graph of consumed power derivative dependence on feed velocity

Simulink Matlab environment analyzes functioning of algorithm described in paper (Pivnyak et al. 2007) (Fig. 4).

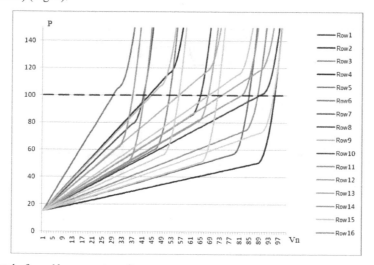

Figure 3. P(Vn) graphs formed by generator unit

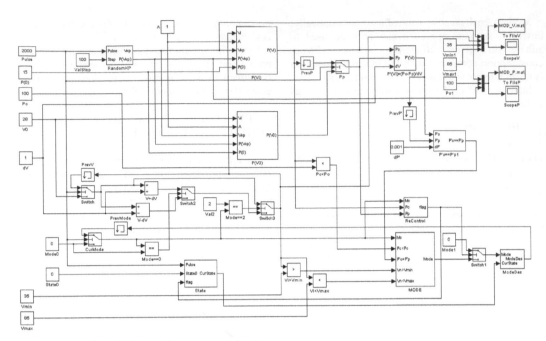

Figure 4. Research model for coal shearer control algorithm

The model helps analyzing the algorithm adaptability for variations in cutting drive operation mode. New Vn.k and P(Vn.k) values are formed by RandomKP unit in which randomizer is implemented for given range. A value experiences one variation per one hundred steps giving the algorithm possibility to adapt to new environment. Figure 3 shows changes in a form of consumed power depending upon cutting drive speed if generator operates. MODE unit explains logics of the algorithm functioning. It determinates transition between modes of speed increase, speed decrease and maintaining cutting drive speed.

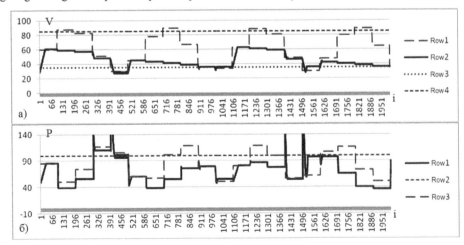

Figure 5. Diagram of step-by-step operation of control algorithm
a) Row 1 is unknown optimum velocity (Vn.k); Row 2 is feed velocity within current step (Vn.i); Row 3 is minimum allowable velocity (Vnmin); Row 4 is maximum allowable velocity (Vnmax)
b) Row 1 is power consumed by a drive (Pi(Vn.i)); Row 2 is maximum allowable power(Po); Row 3 is unknown optimum power (P(Vn.k.))

State block controls accumulation of values required to make decision; ReControl block forms signal for speed change.

Experiment carried out using the model allows analyzing the algorithm behaviour during 2000 step-by-step operations in terms of 20 various values of a coal shearer unknown optimum feed velocity experiencing changes with 100 step regularity. Following limitations are used as input parameters: Vnmin = 35, Vnmax = 85, Po = 100. Initial velocity is V0 = 28, and step of velocity change is dV = 1. The values are defined in relative quantities.

Figure 5 shows diagrams of regulator operation obtained as a result of the experiment.

Consider diagram part during initial 100 steps in detail (Fig. 6).

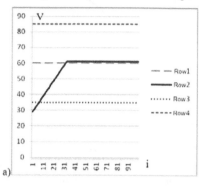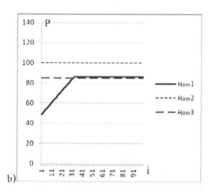

Figure 6. Diagram of control algorithm operation (1-100 steps)
a) Row 1 is Vn.k; Row 2 is Vn.i; Row 3 is Vnmin; Row 4 is Vnmax; b) Row 1 is Pi(Vn.i); Row 2 is Po; Row 3 is P(Vn.k.)

As the Figure explains, generator has formed P(V) curve where Vn.k is 60 (Fig. 6, Row 1) and adequate P(Vn.k) – 85 (Fig. 6, b, Row 3). Limitations are given for minimum allowable (Fig. 6, a, Row 3) and maximum allowable (Fig. 6, a, Row 4) feed drive velocities 35 and 85 respectively. As V0 < Vn.k then the algorithm starts operating in a mode of velocity increase (dV value is equal to 1). Vn.i value reaches the closest to optimum Vn.k within step 33 (Vn.i = 61). Since determined value is within the given limits, transition to a mode of velocity maintenance takes place.

Analysis of diagrams (Fig. 7) shows that the algorithm fails while shifting Vn.k. point to the right that is when current feed velocity should be increased. Regulator identifies the necessity for re-adjusting, makes several steps decreasing current feed velocity which makes it possible to obtain values required to check transition conditions between modes (Fig. 3). This stage is the last one as velocity increase results in the fact that obtained power value turns out to be within linear part of P(V) curve to be a criteria of transition to maintenance mode. Thus, current velocity may be quite less than Vn.k. (Fig. 8).

This effect takes place owing to transition in terms of re-adjusting into a mode of velocity decrease.

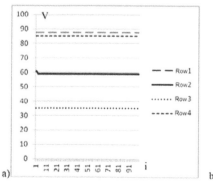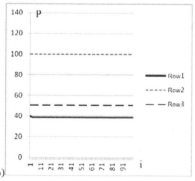

Figure 7. Diagram of control algorithm operation (101-200 steps)
a) Row 1 is Vn.k; Row 2 is Vn.i; Row 3 is Vnmin; Row 4 is Vnmax; b) Row 1 is Pi(Vn.i); Row 2 is Po; Row 3 is P(Vn.k.)

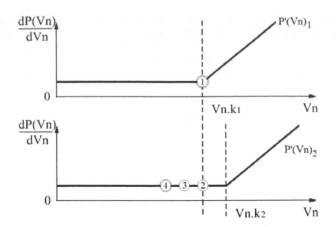

Figure 8. Regulator operation in terms of optimum feed velocity value increase

Produce changes in algorithm which allow transition to velocity increase mode if re-adjustment is required (Fig. 10). Regulator operation starts from accumulating two velocity values and consumed power of cutting drive. Then transition to a mode of velocity increase takes place (M=0). By means of increment to current velocity ΔV value final part of linear area of P(V) curve is determined. Reaching of non-linear area results in the fact that the value of derived of consumed power function varies.

Regulator transfers to a mode of velocity decrease (M=1). In this case return to the closest value being less then Vn.k. is performed and a mode of velocity maintenance is set (M=2). Each step involves check of forming values conformity with the specified limitations.

Consider diagrams being a result of similar experiment mentioned before for modified algorithm (Fig. 9).

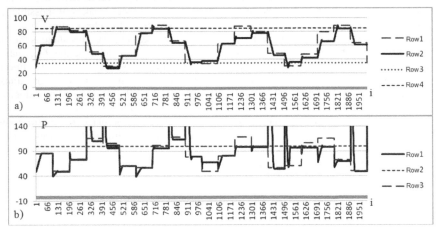

Figure 9. Diagram of step-by-step operation of modified control algorithm
a) Row 1 is Vn.k; Row 2 is Vn.i; Row 3 is Vnmin; Row 4 is Vnmax; b) Row 1 is Pi(Vn.i); Row 2 is Po; Row 3 is P(Vn.k.)

The graphs show that the regulator adjusts dynamically feed velocity to match Vn.k. current value taking into consideration specified limitations. It confirms the algorithm operational improvement. However, detail analysis reveals a number of following disadvantages:

– if Vn.k.>Vnmax, feed velocity first increases up to critical value Vn.k. then decreasing down to

limiting Vnmax, that is upper limitation Vnmax is violated (Fig. 11);

– if velocity decreases by ΔV value, transition to steady mode takes place when feed velocity value is less than critical Vn.k., that is Vn<Vn.k.; hence maximum efficiency cannot be reached (Fig. 12);

– if Vn.k. velocity decrease is reached then the fact that consumed power is out of Po limits is ignored (Fig. 13);

– certain conditions may result in a situation when cyclic increasing/decreasing velocity mode switching occurs, that is steady mode is not available for a regulator (Fig. 14);

– in terms of value Vn.k. change and a corresponding shift of P(Vn) characteristic switching error to steady mode takes place due to changes in values of derived function of coal shearer drive power and when limitations are met (Fig. 15).

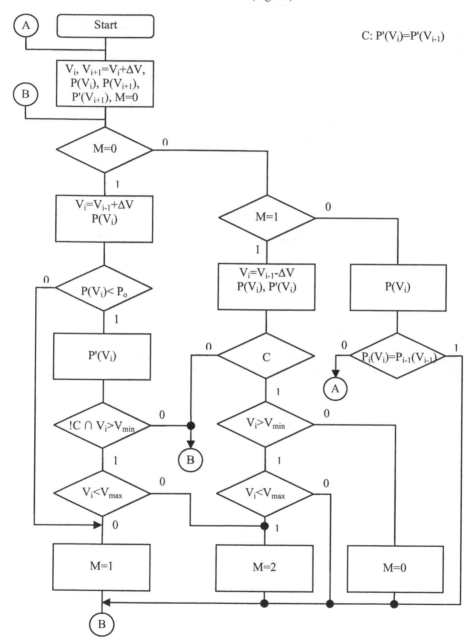

Figure 10. Scheme of coal shearer control algorithm (transition to a mode of velocity increase is by default)

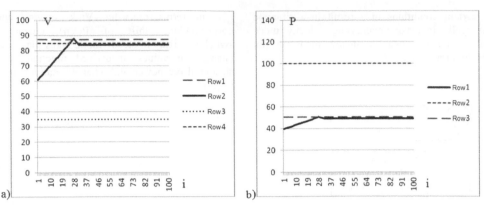

Figure 11. Diagram of step-by-step operation of modified control algorithm (101-200 steps)
a) Row 1 is Vn.k; Row 2 is Vn.i; Row 3 is Vnmin; Row 4 is Vnmax; b) Row 1 is Pi(Vn.i); Row 2 is Po; Row 3 is P(Vn.k.)

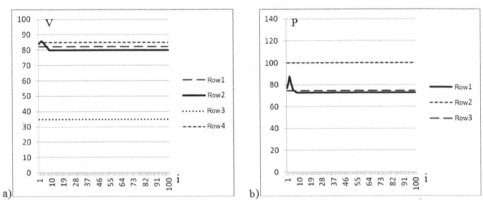

Figure 12. Diagram of step-by-step operation of modified control algorithm (201-300 steps)
a) Row 1 is Vn.k; Row 2 is Vn.i; Row 3 is Vnmin; Row 4 is Vnmax; b) Row 1 is Pi(Vn.i); Row 2 is Po; Row 3 is P(Vn.k.)

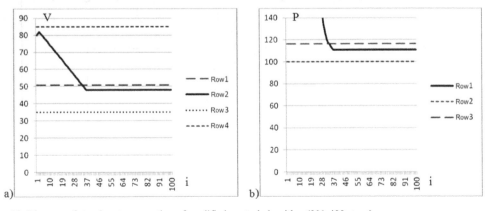

Figure 13. Diagram of step-by-step operation of modified control algorithm (301-400 steps)
a) Row 1 is Vn.k; Row 2 is Vn.i; Row 3 is Vnmin; Row 4 is Vnmax; b) Row 1 is Pi(Vn.i); Row 2 is Po; Row 3 is P(Vn.k.)

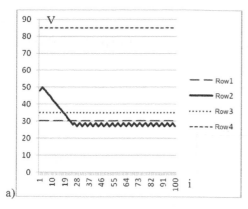 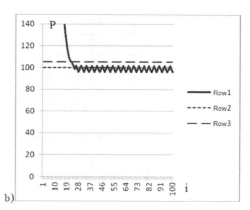

Figure 14. Diagram of step-by-step operation of modified control algorithm (401-500 steps)
a) Row 1 is Vn.k; Row 2 is Vn.i; Row 3 is Vnmin; Row 4 is Vnmax; b) Row 1 is Pi(Vn.i); Row 2 is Po; Row 3 is P(Vn.k.)

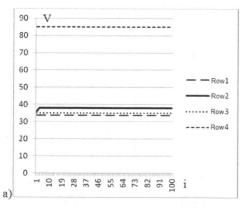 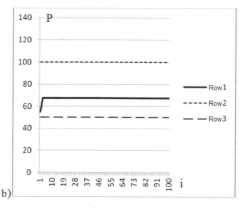

Figure 15. Diagram of step-by-step operation of modified control algorithm (1001-1100 steps)
a) Row 1 is Vn.k; Row 2 is Vn.i; Row 3 is Vnmin; Row 4 is Vnmax; b) Row 1 is Pi(Vn.i); Row 2 is Po; Row 3 is P(Vn.k.)

In addition transition in terms of Vn.k. point displacement from maintenance mode to a mode of velocity increase may influence negatively on a coal shearer cutting drive as Vnmax or Po may be reached before accumulation of values required for regulator.

To correct above-mentioned disadvantages a new algorithm of regulator operation has been developed (Fig. 16).

Control process may be in one of three modes (M): velocity increase (0), velocity decrease (1), and current velocity maintaining (2). Transitions are performed in terms of checking the conditions: limitations on power (Pi(Vi)<Po), limitations on minimum (Vi≥Vnmin) and maximum velocities (Vi≤Vnmax), reaching optimum velocity (|Pi'(Vi)|=|Pi'-1(Vi-1)|). Besides, previous mode (PM) and a state inside a mode (S) are controlled providing accumulation of values required for conventional transitions.

Experiment carried out using the model has allowed obtaining diagrams shown in Fig. 17.

94

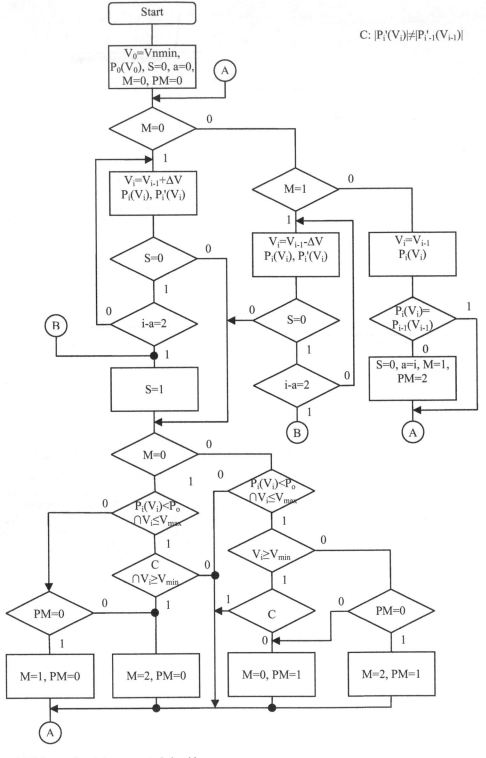

Figure 16. Scheme of coal shearer control algorithm

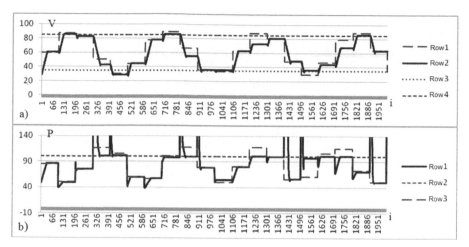

a)

b)

Figure 17. Diagram of step-by-step operation of control algorithm
a) Row 1 is Vn.k; Row 2 is Vn.i; Row 3 is Vnmin; Row 4 is Vnmax ;b) Row 1 is Pi(Vn.i); Row 2 is Po; Row 3 is P(Vn.k.)

Diagrams of Fig. 18 compare operations of new algorithm with that analyzed before.

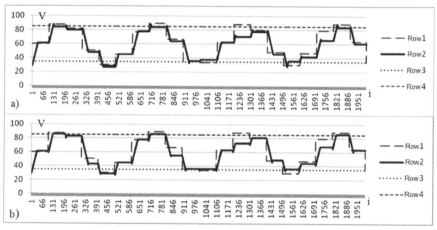

a)

b)

Figure 18. Diagram of step-by-step operation of control algorithm a) modified, b) new algorithm
Row 1 is Vn.k; Row 2 is Vn.i; Row 3 is Vnmin; Row 4 is Vnmax

Consider in detail (101-200), (201-300), (301-400), (401-500), and (1001-1100) intervals within which disadvantages have been marked (Fig. 19-23).

Analysis of the diagrams allows drawing following conclusions:

– mode of velocity increase if Vn.k.>Vnmax can no longer result in violation of upper limitation Vnmax (Fig. 19);

– transition from decreasing mode to maintenance mode takes place at reaching Vn.k. (Fig. 20);

– limitation of consumed power Po is strictly met (Fig. 21);

– steady operation mode can be implemented for the regulator under any conditions (Fig. 22);

– in terms of changes in Vn.k. value and corresponding shift of P(Vn) characteristics no switching error into steady mode occurs (Fig. 23);

– in terms of defaul re-adjustment regulator transfers to a mode of velocity decrease.

96

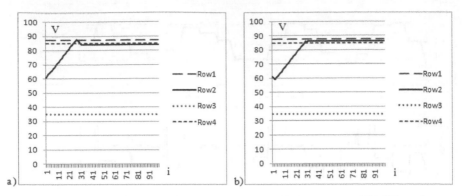

Figure 19. Comparison of control algorithm operation diagrams: a – modified, b – new one (101-200 steps)
Row 1 is Vn.k; Row 2 is Vn.i; Row 3 is Vnmin; Row 4 is Vnmax

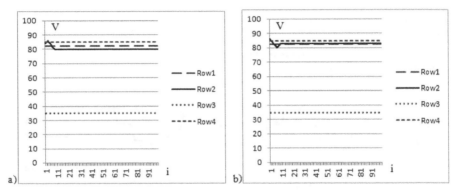

Figure 20. Comparison of control algorithm operation diagrams: a – modified, b – new one (201-300 steps)
Row 1 is Vn.k; Row 2 is Vn.i; Row 3 is Vnmin; Row 4 is Vnmax

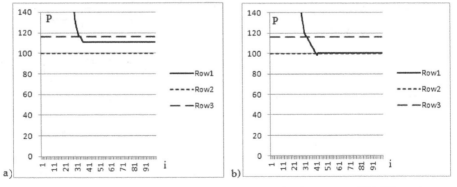

Figure 21. Comparison of control algorithm operation diagrams: a – modified, b – new one (301-400 steps)
Row 1 is Pi(Vn.i); Row 2 is Po; Row 3 is P(Vn.k.)

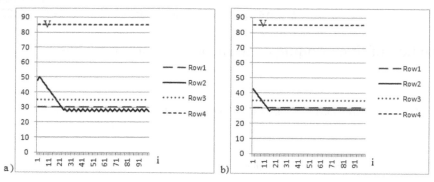

Figure 22. Comparison of control algorithm operation diagrams: a – modified, b – new one (401-500 steps)
Row 1 is Vn.k; Row 2 is Vn.i; Row 3 is Vnmin; Row 4 is Vnmax

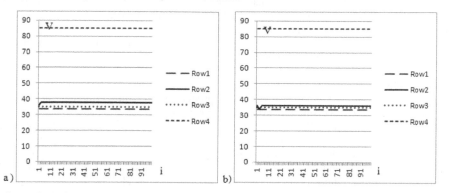

Figure 23. Comparison of control algorithm operation diagrams: a – modified, b – new one (1001-1100 steps)
Row 1 is Vn.k; Row 2 is Vn.i; Row 3 is Vnmin; Row 4 is Vnmax

4 CONCLUSIONS

1. Algorithm of determining P(Vn) (Vn.k) characteristic bending point on the basis of readings of sensor of power consumed by feed drive may be used to develop a regulator to stabilize main motor load.

2. Feature of the approach being the basis for the algorithm is step-by-step interval determination within which linear part of P(Vn) function transforms into non-linear one. Moreover, quickness and accuracy of determining optimum feed velocity for current operation mode depend directly on step value ΔV. When operation mode varies re-adjustment starts after three intermediate values have been accumulated; thus, periodicity of power sensor monitoring effects directly the regulator operation.

3. The algorithm application will help regulator of load stabilization for cutting organ of main motor maintain feed velocity value involving Vnmin, Vnmax, and Po limitations.

REFERENCES

Pivnyak G.G., Protsenko S.N., Stadnik N.I., Tkachov V.V., 2007. *Decentralized control. Monograph.* (in Russian). Dnipropetrovs'k.: NMU: 107.

Power Engineering, Control and Information Technologies in Geotechnical Systems – Pivnyak, Beshta
& Alekseyev (eds)
© 2015 Taylor & Francis Group, London, ISBN 978-1-138-02804-3

Optimization of fine grinding on the acoustic monitoring basis

N. Pryadko
State Higher Educational Institution "National Mining University", Dnipropetrovs'k, Ukraine

ABSTRACT: The optimization purpose of thin grinding process is mill productivity increase and power consumption decrease with necessary requirements of grinding product dispersion. Several directions of grinding process optimization are developed on the basis of the signal analysis results of grinding and classification zone: the continuous the maximal amplitude control during grinding; the energy analysis of signals; the result analysis of spatial model visualization and model on the circuits Markov basis; the Wavelet-analysis, Hilbert – Huang transformations and neural network signal analysis; the technological - acoustic criteria and factor control. The applications of the developed information thin grinding technology are considered for laboratory and industrial mills.

1 INTRODUCTION

In technology of mineral processing fine grinding are the most power-intensive. That's why the problem of an establishment of optimum grinding conditions (Pivnyak 2014) is actual one. Last discoveries in physics of durability and destruction gave an impulse for researching of the grinding physical mechanism. The new approach to research of grinding effects during destruction and grinding is based on using of the principle of power spectra similarity at destruction of laboratory samples and mountain ranges, laws of explosive (relaxation) self-oscillations of the condensed substance in the predestruction theory, the conclusion about a determining role of damage accumulation in models of rock volumetric - deformed conditions (Kuksenko 2005).

According to (Pryadko 2008), the method of acoustic emission gives the information about all stages of the kinetic destruction process (flaw formation, crushing, grinding) and expected effects of grinding. The main acoustic characteristics in the forecast of grinding effects are specific acoustic radiation, amplitudes and activity kinetics at a stage of over-limited deformation. In (Pilov 2009) it is shown, that there is a correlation between acoustic signal amplitude and the destruction size at compression of rock laboratory samples and impacts of particles during jet grinding. Thus, the particles' size change during grinding causes transformation of amplitude distributions.

Checking of theoretical conclusions about applicability of an acoustic method for monitoring grinding process was carried out on a jet mill which

mechanism provides technology requirements of producing of fine grinding activated mineral powders.

Nowadays the problems of an establishment and maintenance of jet mill optimum operating mode for achievement of required ready product quality at the minimal expenses of energy are studied. Well-known technical decisions of jet grinding optimization by the control of temperature or depression over a jet mill path have a time delay and significant error in an estimation of grinding parameters: mill productivity and power consumption, product fineness.

2 FORMULATION THE PROBLEM

There are some ways of the automatic control and regulation of grinding process on the basis of active capacity gauging of the mill engine and productivity of mill loading, etc. However these ways don't allow making operative decision on a mode or technological parameters change without a stop of grinding process.

3 MATERIALS FOR RESEARCH

In (Pryadko 2012; Pryadko 2013; Gorobets 2013; Pilov 2013) acoustic monitoring of a jet mill is approved for the control of jet loading by material and crushed particles fineness. The grinding mode was estimated by quantity of the ground product, mill productivity and product grinding. Tests have shown that mill productivity deviates from a maximum level in both jets over and under load. Measurement of a acoustic activity level of a grinding zone was carried out with the help of the acoustic gauge connected with a brass wave guide, set inside mill chambers. For quality control of the ground product the second wave guide and the gauge were in a zone behind the classifier. Gauges through the analog-digital converter are connected with a computer on which the analysis of signals and their preservation are realized.

On Fig. 1 the jet mill circuit with acoustic monitoring system is shown.

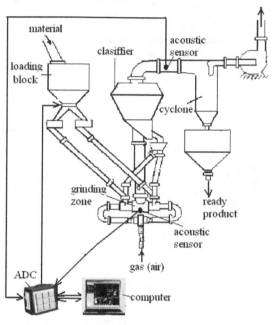

Figure 1. Jet mill circuit with system of acoustic monitoring process

As a result of grinding of various physical properties materials and the set technological parameters connections of acoustic signal parameters and technological grinding modes are established. The database is created which includes the material characteristics, grinding and size structure of grinding products, and researched parameter record.

Application of acoustic monitoring parameters for optimization of jet grinding process has the following advantages:

- the continuous control of process and change of mill loading value realizes without a stop of technological process,

- according to connection of acoustic and process technological parameters the recommendation on management of grinding process are given out in current time.

- according to signals of a classification zone there is a continuous quality control of a received product dispersion). If it's need recommendations on change of a classification mode are given out.

In this work management of grinding process with the purpose of mill productivity increase or for decrease of mill power consumption at execution of necessary requirements to grinding product dispersion is offered to carry out on the basis of acoustic monitoring.

On the basis of acoustic monitoring results some directions of grinding process optimization are developed:

1. The continuous control of the maximal amplitude of signals during grinding;

2. The control of grinding process over the established acoustic – technological criteria and factors;

3. Quality control of the crushed product;

4. The continuous power analysis of acoustic signals;

5. Wavelet-analysis system of the acoustic signals in grinding zone;

6. Neural network analysis of grinding modes (Pryadko 2010);

7. The surface analysis of density functions of signal amplitudes probability distribution in characteristic zones;

8. Modeling grinding process on the basis of Markov circuits.

100

Each of these directions has its advantages, disadvantages and field of application about which it is told below.

The continuous control of the signal maximal amplitude during grinding.

On the basis of experimental results dependences of acoustic signal amplitudes on the crushed particle sizes, and acoustic signal activity $\dot{N}$ from a jet loading mode have been established. The experimental data confirm dependences of technological and acoustic parameters as $\dot{N} = f(G, P, t)$ and $A = f(P, n)$, where P, t - pressure and temperature of the energy carrier (air), G - mill productivity, n - number of qualifier revolutions per minute, A, N - amplitude and activity of acoustic signals.

On Fig. 2 it is shown kinetics amplitudes (a) and activity (b) of acoustic signals on various modes of chamotte grinding.

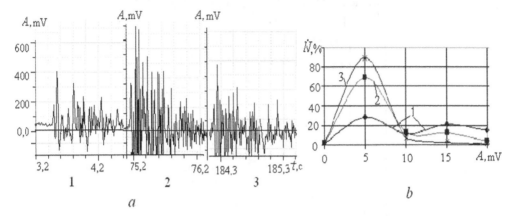

Figure 2. Kinetics amplitudes (a) and activity (b) of acoustic signals in a grinding zone at various process stages (1 - loading, 2 - an operating conditions, 3 - unloading)

The result analysis of loose materials jet grinding has allowed to lay down the following laws of jet mill acoustic signal:

- The amplitude of acoustic signals and its size distribution characterize a degree of jet loading by material;
- Overdose jet loading by material (a mode of loading and an overload) is accompanied by reduction of acoustic signal amplitude that specifies decrease in grinding dynamics;
- The size of signal amplitude in characteristic not optimum jet conditions (loading and overloading) has significant distinctions (up to some orders), that allows to consider using of acoustic signal monitoring for jet grinding process regulation as perspective one.

On this basis it is offered to carry out optimization of a jet mill functioning by the continuous acoustic activity control of a grinding zone and to operate loading of crushed material required quantity according to optimum jet fullness by a firm phase (Pryadko 2013).

The grinding process control over the revealed acoustic - technological criteria and factors.

Taking into account connection of acoustic and technological parameters, criteria of optimum work of industrial and laboratory mills and their connections with mill productivity and crushed powder dispersion are determined and proved.

During monitoring grinding acoustic signal activity and amplitude spectra are measured, maximal amplitudes A_{max} and the general signal number $\dot{N}_{\Sigma}$ for the chosen time intervals (Δt, s, ms) are determined. By measurement results current values of the optimality criterion $K_{opt} = A_{max} \cdot \dot{N}_{\Sigma}$ are calculated. If operation of ready product weighing and the change control of mill productivity are available an opportunity of acoustic signal monitoring is extending. Simultaneously conducting monitoring other suggested criteria K_{ef}, K_c are calculated.

On fig. 3 the schematic behavior image of jet grinding acoustic ($\dot{N} = N/\tau$, A_{max}) and technological parameters is given: mill productivity, product dispersion, efficiency criteria $K_{ef} = G/\lg \dot{N}$ (g/imp) and material circulation $K_c = (N/N_{feed})$. Process is examined at constant parameters of the energy carrier ($P = 0,3$ MPa) and a classification mode ($n = 600$ min^{-1}). The modes describing various conditions of jet loading by a material are designated by the following intervals: $t_1 t_2$ - loading of a material in weight m_1 and an output on an operating conditions of grinding; t_2t_3 - an optimum grinding mode with the best productivity (during the period t_{ef}); t_3t_4 - loading of a material in weight $m_2 > m_1$ with some overload of jets; t_4t_5 - an operating conditions; t_5t_6 – jet unloading.

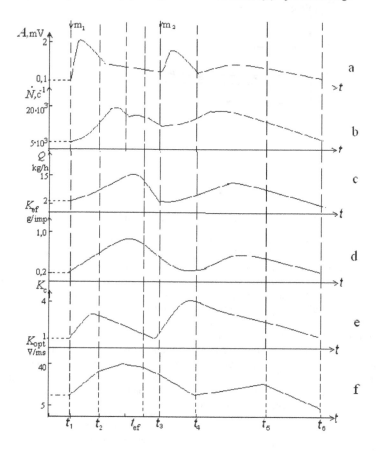

Figure 3. The schematic image of informative acoustic and technological parameters of jet grinding process.

The set acoustic parameters of efficiency K_{ef} and circulation K_c (fig. 3c, d) reflect different aspects of grinding process. Jet grinding efficiency is estimated by efficiency criterion $K_{ef} = G/\lg \dot{N}$ (g/imp), describing a degree of grinding process approach to optimum conditions. Experiments have shown, that K_{ef} value should be more than 0,4 for exception of ineffective work of a mill. In an optimum grinding mode parameter K_{ef} is equal to 0,8-0,9. The circulation factor $K_c \approx N/N_{feed}$ is estimated as a ratio of the current account N of acoustic signal to initial N_{feed} ones at the moment of loading and depends on a degree of jet filling by material. In optimum mill work conditions this factor is about unit, whereas in jet overflow conditions its value rise to 2 - 4, and at excessive unloading it is less than 1.

The suggested optimization criteria K_{opt} is calculated as product of maximal amplitude A_{max} on the general number of signals or number of signals with the maximal amplitude, i.e. K_{opt} $(\dot{N}_\Sigma) = A_{Max} \cdot \dot{N}_\Sigma$. From physics positions of destruction this criterion characterizes acoustic effect of kinetic energy transformation of the particle accelerated by jets in acoustic flaw formation energy at destruction of particles by impacts and, thus, the value of the offered criterion is theoretically proportional particle destructions rating during jet grinding. The analysis of acoustic monitoring data of jet grinding process of various loose materials (quartz sand, slag, chamotte, zircon, coal) has allowed to designate borders (allowable and inadmissible from the positions of a process optimality) changes K_{opt} for one grinding cycle

102

(loading of one portion, grinding, jet unloading, a product unloading). The Fig. 4 illustrates kinetics of acoustic optimum criterion for laboratory (1) and industrial (2) mill action for zircon grinding.

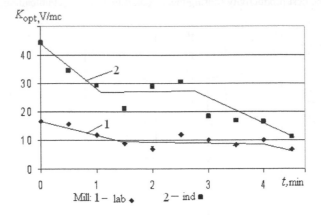

Figure 4. Kinetics of acoustic optimum criterion for laboratory (1) and industrial (2) mill action for zircon grinding

These In the limits of one grinding cycle the tendency of value K_{opt} reduction is observed which is explained by the current jet unloading from a firm phase during ready product removal to a cyclone, and, an inclination corner of this dependence is the much greater for industrial grinding conditions. Hence, it's extremely important to provide a duly loading of new portions of an initial material with carrying out of continuous acoustic monitoring that allows to have a optimality criterion K_{opt} value at a level above its allowable value of criterion

Thus, a jet grinding process optimization with the help of acoustic - technological criteria bases on a law

$$\dot{N}_{A_{max}} = f(G, S_{sp}, K_{ef}, K_{opt}, K_c),$$

which characterizes activity $\dot{N}$ changes of acoustic signal with the maximal amplitude depending on technological parameters of quantity (G) and crushed product qualities (Ssp).

On the comparison basis of the criteria current value with required one it is judged a jet condition, efficiency and power consumption of process, ground product quality and the further regulation actions of technology. If the criteria current value chosen for the jet loading control by material, is beyond allowable one, the signal is made either on a mill stop, or on additional, next loading of an initial material portion (in case of grinding continuation). At inadmissible change of the criteria which control product quality, the signal on change of a classification mode is given, and if it's necessary updating of energy carrier parameters is made.

The crushed product quality control.

Results of researches show, that accumulation process in spectra of acoustic signals with small amplitude values (about 5-20 mV) characterizes the primary contents in jet small size particles, i.e. material dispersion effects are prevailed in grinding kinetics. On the experimental data of laboratory jet grinding connections of acoustic parameters with crushed products specific surface S_{sp}. are established.

It is set, that increase of crushed powder dispersion is accompanied by reduction of characteristic (maximal) amplitude A_{max} and growth of individual share of signals with small amplitude. For materials with density of 2,8-3,0 g/sm^3 connection of acoustic parameters A_{Max}, and $\dot{N}(A_{-40})$, % with a dispersion parameter S_{sp}. (m^2/g) is described with the equations:

$$\dot{N}(A_{-40}) = 16 \cdot S_{sp} + 87$$

and

$$A_{-40} = 1749 - 2172 \cdot S_{sp},$$

with approximation reliability $R = 0,87$ and $R = 0,93$, accordingly.

The analysis of classification zone signals allows controlling crushed product quality. On fig. 5 signal records are shown at a different classification mode.

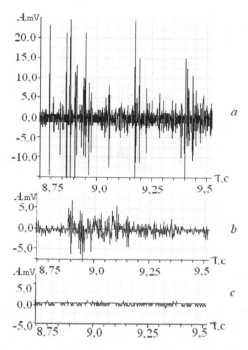

Figure 5. Records of signals on an output from the qualifier at different classification modes: $a-n=0$; b –500, c – 1000 min^{-1}

Without classification *(a)* the coarse product is produced, signal amplitudes are larger than on the order the signals which have been written down at the working qualifier (b, c). Such quality monitoring allows to notice large particles occurrence in a ready product in due time and to avoid a product spoilage.

The continuous energy analysis of acoustic signals

Researches of informative frequencies of acoustic signals were carried out at grinding various materials. In particular, on fig. 7 energy change of different frequency signals is shown for quartz sand jet grinding. Signals in intervals duration $\Delta t = 1c$ for one cycle of mill loading are chosen. Signal registration frequency is 400 kHz, therefore the informative frequencies spectrum has been determined as $65\ kHz \le f \le 125\ kHz$. In an operating grinding conditions $1c \le t \le 3{,}5c$ signal amplitudes are larger, than at the initial loading stage and, that is more important, it's larger than at the final grinding stage, i.e. in unloading mill. Especially it is typical for signals with the frequency getting in an informative range (Pryadko 2012).

On the basis of the examined connection of change laws of jet grinding technological parameters and amplitude - frequency characteristics of the acoustic monitoring data the process control algorithm is offered.

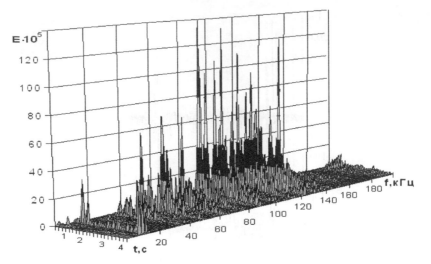

Figure 6. Distribution of signal energy on frequency strips during quartz sand grinding the beginning of material loading is at $0 \le t \le 1$, min; an operating grinding conditions is at $1 \le t \le 3,5$, min; insufficient mill loading is at $3,5 \le t \le 4$, min.

On the basis of the examined connection of change laws of jet grinding technological parameters and amplitude - frequency characteristics of the acoustic monitoring data the process control algorithm is offered.

The surface analysis of density functions of signal amplitudes probability distribution in characteristic zones

The developed visualization algorithms allow revealing a new recognition way of mill loading degree according to acoustic signals.

Research of two researching zone signals (grinding and classification) simultaneously at various grinding stages has shown dependence of acoustic parameters on a degree of jet loading by material. Filling of a database by acoustic signal monitoring of grinding process and development of information grinding system have enabled to visualize results of the acoustic signal amplitudes analysis. As a visualization problem of mill acoustic signal monitoring it is understood the following. According to monitoring of such kind $\{t_i, A_i^m, A_i^k; i = 1,...N\}$, where N – amount of signal records in a second ($i = 1...N$); the data pair defines kinetics of mill signal amplitudes ($\{t_i, A_i^m; i = 1,...N\}$) and signal amplitudes after the qualifier ($\{t_i, A_i^k; i = 1,...N\}$). By acoustic monitoring results the density functions of signal amplitudes probability is found. It is function $p(A^m, A^k)$, which realization is observable amplitude files.

Realization of a visualization task is carried out on a basis of the regularized monitoring mill data after noise suppression by the wavelet-analysis with the help of second order Dobeshy Wavelet. Restoring function of probabilities distribution density is Gauss function like $p(A^m, A^k)$. The flat and volumetric picture analysis of probabilities distribution density function $p(A^m, A^k)$ of grinding acoustic signals of various density materials allows to judge mill congestion by material.

On fig. 7 the function of probabilities distribution density for two mill loading modes are shown. At the acoustic signals analysis of quartz grinding process in mill acting mode the section area of the function schedule $p(A^m, A^k)$ by a coordinate plane (0xy) twice larger than the similar area in an unloading mode (Fig. 7c and 7d). At the same time in volumetric representation (Fig. 7a and 7b) function peak in operating grinding conditions is three times higher than corresponding representation of unloading mode signals.

105

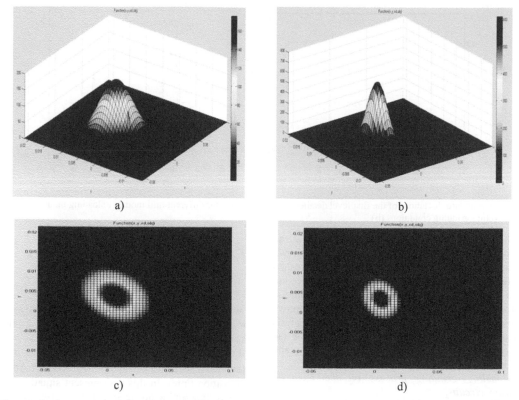

Figure 7. Surface analysis of density functions of signal amplitudes probability distribution and its projection in characteristic zones: operating conditions of grinding (a, c) and empty mill (b, d)

The suggested graphic addition to the acoustic signal data analysis as surface of density functions of signal amplitudes probability distribution and its projection to a plane give the additional information about signal behavior features, helps in condition definition of mill congestion. Presentation of schedules gives more complete recognition of mill congestion, disorder of signal amplitude values, a prevalence of those or other signals.

Wavelet-analysis system of the acoustic signals in grinding zone.

The change law analysis of jet grinding technological parameters and amplitude -frequency signal characteristics of a grinding zone has allowed developing system of the Wavelet - analysis of jet mill acoustic signals (Pryadko, Gorobets, Levchenko & Verhorobina 2013).

The researches of the jet grinding acoustic monitoring data show, that Wavelet - factors of two various modes (optimum and jet insufficient loading by a material) differ on the order; therefore they well enough determine a mill filling degree.

2. The Wavelet-analysis of a jet mill signals at a unloading stage by grinding materials of various physical properties has shown, that standard deviations of the first level detailed Wavelet - factor practically do not differ, that allows developing of mill loading control algorithm for jet grinding of various sized loose materials.

At the Wavelet-analysis of grinding operating conditions of various density materials (gas coal $\rho = 1,4$ g/m^3 and quartz sand $\rho = 2,65$ g/m^3) and different dispersion of the crushed product, various estimation value of first level detailed Wavelet-factor are received. On fig. 8 standard deviations of mill unloading mode (a) and an operating grinding conditions (b) of quartz sand are shown

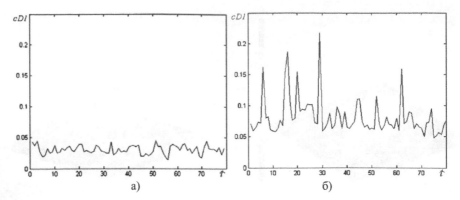

Figure 8. Standard deviations of the first level detailed Wavelet - factor for different mill modes: unloading mode (a) and an operating grinding conditions (b)

Investigation of grinding zone signals at an optimum grinding mode (Mihalev 2013) has shown dependence of Wavelet - factors from crushed material properties that allow controlling dispersion of a received product during grinding. According to carried out research results it is possible to approve about applicability of the developed analysis technique of jet grinding acoustic signals for the process control, increase of its efficiency and quality control of the crushed product.

Modeling grinding process on the basis of Markov circuits.

Modeling is carried out on the basis of the system approach with use of cellular model. At modeling grinding process in the jet mill with a periodic mode (periodic material loading), all mill is accepted as elementary volume of modeling.

For creation cellular model all acoustic signal range is broken on sub-band (sub-range) of signals with the average amplitude sizes A_i, $i = 1,2,..m$, where $i = 1$ corresponds A_{max}. Thus, all acoustic signals which are written down during monitoring, are broken on m cells according to their amplitude value. The number of corresponding amplitude signals determines an elementary cell condition. The amount of acoustic signals can be submitted by a column vector $N = (n_i)$, $i = 1,...m$, where n_i - number of acoustic signals in i cell, i.e. number of the signals with amplitude A_i.

Each condition is characterized by the certain probability which can be considered as a share of signals with corresponding amplitude A_i, and also as dimensionless number of signals. Transition to the normalized distribution of acoustic signals number in each cell at any splitting into classes is realized under the formula:

$$n_i^k = \frac{n_i}{\sum_1^m n_i} = \frac{n_i}{N},$$

where N – number of registered acoustic signals for an monitoring interval, i - a cell number, k - the discrete time moment of monitoring.

For modeling jet grinding process kinetics it is examined a sequence of small time intervals (Δt is a transition time). In this case present situations of process time and acoustic signal monitoring record can be determined as $t_k = k\Delta t$, where an integer number $k = 1,2...$ is the transition one. It can be considered as a discrete present time moment.

For k transition condition vector N^k passes into N^{k+1}. Then amplitude change of written down signal during grinding for a time step of transition in cell model is calculated from matrix equality $A^{k+1} = GA^k$. The transition number k is calculated from expression $k = t/\Delta t$, where t is discrete duration of material stay in a mill. For the description grinding kinetics it is important to determine selective and distributive function.

In the theory of signals definition of specific energy and average specific capacity of the signals is used which have been written down for the given time interval t as $E_A = \int_0^t s^2(t)dt$, $P_A = \frac{1}{t}\int_0^t s^2 dt$.

Destruction of particles at which there appear signals with amplitude A_i, occurs when signals possess specific capacity of energy $P_A = \frac{1}{t}\sum_{i=1}^m n_i A_i^2$.

Using expression for capacity of signals, we receive

107

expression for definition of grinding selective and distributive function $S_i = \dfrac{1 - P_{i,j}}{\Delta t}$, $b_{k,i} = \dfrac{P_{k,i}}{S_i \Delta t}$ with the help of which a grinding matrix can be found. The received approbation results of mathematical model on quartz sand grinding are corresponded to the earlier established low of connection acoustic signals and ground product dispersion at jet grinding. Thus, the mode and grinding product dispersion can be identified on the basis of created sell models of jet grinding acoustic monitoring.

4 CONCLUSIONS

Maintenance of optimum parameters during jet mill operating is offered to be realized by continuous monitoring of a grinding zone acoustic activity on the basis of the following considered jet grinding optimization methods: continuous monitoring of the maximal amplitude and activity of jet mill acoustic signals (Pilov, Gorobets, Pryadko 2014), the control of circulation and efficiency factors, maintenance at a necessary level the optimality criterion value; graphic visualization of the signal energy analysis of a grinding zone and a surface analysis of density functions of signal amplitudes probability distribution in characteristic zones (Pryadko, Bulana 2012); the Wavelet-analysis and Neural network analysis of grinding modes, modeling of grinding process on the basis of Markov's circuits.

All approaches are based on the results of process acoustic monitoring and allow operating grinding material loading on the basis of search of an optimum jet filling by a firm phase. The found laws and the received experimental data have allowed developing information optimization technology of jet grinding process.

REFERENCES

Pivnyak G.G., Pilov P.I., Pryadko N.S. 2014. *Decrease of Power Consumption in Fine Grinding of Minerals*. Springer: Mine Planning and Equipment Selection: 1069 -1079.

Kukcenko V.S. 2005. *Diagnostics and forecast of large-scale object destruction* (in Russian). Physics of solid bodies, Issue 47, Vol. 5: 788 – 794.

Pilov P.I., Gorobets L.J., Bovenko V.N., Pryadko N.S. 2008. *An acoustic monitoring of the sizes changes of grinded particles* (in Russian). Dnipropetrovs'k: Scientific herald of NMU, Issue 6: 23 – 26.

Pilov P.I., Gorobets L.J., Pryadko N.S. 2009. *Research of acoustic monitoring regularities in a jet grinding process.* Polish Academy of Sciences: Mining Sciences Archives, Issue 54, Vol. 4: 841 –848.

Pryadko N.S. 2012. *Acoustic-emission monitoring of jet grinding process* (in Russian). Technical diagnostics and nondestructive check, Issue 6: 46 – 52.

Pryadko N.S. 2013. *Acoustic research of jet grinding* (in Russian). Saarbrucken: LAP LAMBERT Academic Publishing.– OmniScriptum GmbH&Co.Kg.: 172.

Gorobets L.J., Bovenko V.N., Pryadko N.S. 2013. *Acoustic method of grinding process studying* (in Russian). Saint Petersburg: Dressing of ores, Issue 3:18-24.

Pilov P.I., Gorobets L.J., Pryadko N.S., Krasnoper V.P. 2013. *Acoustic principles of jet grinding* (in Russian). Dnipropetrovs'k: Dressing of minerals, Issue 54 (95): 30- 37.

Pryadko N.S., Gorobets L.J., Levchenko K.A. & Verhorobina I.V. 2013. *Principles of fine grinding acoustic optimization* (in Russian). Kharkiv: NTU "KhPI": Scientific Transactions of the National Technical University "Kharkiv Polytechnic Institute", Issue 27: 71-81.

Pryadko N.S., Bulana T.M., Gorobets L.J., Sobolevskaya U.G., Sirotkina N.P. 2010. *Informational technology of getting fine dispersion material by jet grinding* (in Russian). Dnipropetrovs'k: System technologies, Issue 3 (58): 40-46.

Mikhalyov A., Pryadko N., Suhomlin R. 2013. *Fine grinding analysis on bases of Hilbert-Huang and Wavelet transformation of acoustic signals.* Moscow: MUSAm: The materials of the IX International Congress of dressers of CIS: 409-412.

Mikhalyov A., Pryadko N., Suhomlin R., Kotyra A. 2013. *Application of wavelet transform in analysis of jet grinding process.* Lublin: Elektronika, Issue 8: 20-22.

Pilov P.I., Gorobets L,J., Pryadko N.S. 2014. *Jet grinding monitoring method and jet mill* (in Ukrainian). Patent of Ukraine № 104427, B 02C 25/00 /published 10.02.2014, bulletin № 3, application № a201016004 published 10.07.2012, bulletin № 13.

Pryadko N.S., Bulana T.M. 2012. *Jet grinding monitoring method* (in Ukrainian). Patent of Ukraine № 73291 B 02C 25/00 /published 25.09.2012, bulletin № 18, application № u201114725 published 12.12.2011

Flexible couplings with rubber-cord shells in heavy machinery drives

B. Vinogradov & A. Khristenko
State Higher Education Institution "Ukrainian State Chemical Technology University", Dnipropetrovs'k, Ukraine

ABSTRACT: The effectiveness of flexible couplings with rubber-cord shells in tumbling mill drives has been studied. It has been revealed that along with high compensatory characteristics, these couplings can reduce the dynamic loads and evenly distribute the load between the transmission lines in the twin-motor drives.

1 INTRODUCTION

Flexible couplings for heavy machinery e.g. large tumbling mills are designed for kinematic and power shaft-to-shaft connections, limitation of dynamic loads and compensation of axial and angular misalignments of shafts. In dealing with these problems, shaft couplings equipped with rubber-cord shells as elastic elements are of interest.

Currently, rubber-cord shells are widely used as air springs (Borodulin, Brener, Grechishchev et al. 1976) in vibration isolators (Pastukhov, Rosin 1977). Studies of viscoelastic characteristics of the systems using rubber-cord shells have shown their efficiency in dynamic load limiting and resonance oscillation damping (Vinogradov 2011).

The aim of the article is to assess the effectiveness of flexible rubber-cord shell couplings in heavy machinery drives on an example of tumbling mills.

2 MATERIALS FOR RESEARCH

As an example, we will consider Ya-300 rubber-cord shell. At an internal pressure of $p = 0.51$ MPa, the load-carrying capacity of a pneumatic element is $Q = 147100$ N, and with a reference height of $H = 165$ mm, a compression and rebound stroke is $Z = \pm 40$ mm. Five pneumatic elements being installed in the coupling (Fig. 1a), a torque of $M_{torq} = 509$ kN · m can be transmitted, while eight pneumatic elements allow transmitting a torque of $M_{torq} = 814$ kN · m (Fig. 1c). It should be noted that the largest diameter will be $D_2 = 2200$ мм, while the diameter of the pneumatic elements installed will be $D_1 = 1384$ mm. To reduce the coupling stiffness and increase the

coupling movement, the pneumatic elements can be installed in series (Fig. 1b). The coupling stiffness can further be reduced by connecting the pneumatic elements to an additional volume.

To improve the dissipative properties of the coupling, rubber-cord shells can be filled with fluid and connected by a throttle to a hydropneumatic accumulator which can be mounted directly on the shaft (Vinogradov 2011).

Work area of the coupling viscoelastic characteristics is described by the following expression (Vinogradov 2004):

$$M(\alpha) = n_{\text{ПБ}}\, R_{\text{M}}\, S(\alpha) \times$$

$$\times \left[\frac{\rho\left(S(\alpha)\right)^2 R_{\text{M}}^2}{2\,\mu_1^2\, f^2}\left(\frac{d\alpha}{dt}\right)^2 + (p_a + p_{uo}) \times \left(\frac{V_{zo}}{V_{zo} - R_{\text{M}}\int_0^\alpha S(\alpha)d\alpha}\right)^n - p_a \right] \tag{1}$$

$$S(\alpha) = S_0 \cdot \left(1 + k_a \alpha R_{\text{M}}\right)$$

where α is the coupling torsion angle; $M(\alpha)$ is the moment reacted by the pneumatic coupling; $S(\alpha)$ is the effective area of a rubber-cord shell; S_0 is the initial effective area of a rubber-cord shell at rated load; k_a is the approximation factor; V_{zo} is the volume of gas in a pneumatic element (in a rubber-cord shell or a pneumatic chamber of a hydropneumatic accumulator); p_a, p_{uo} are the atmospheric and excess pressure within the fluid and gas; ρ is the fluid density; μ_1 is a coefficient of the fluid flow rate through the throttling orifice; f is the throttle flow area; n is the polytrope coefficient

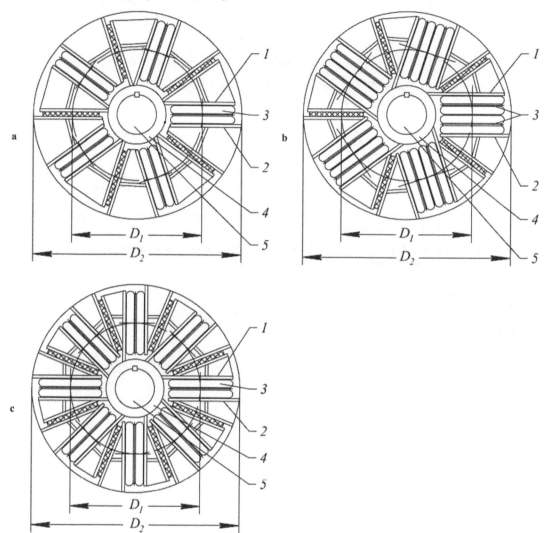

Figure 1. Flexible couplings with rubber-cord shell Ya-300: a, b – torque M_{torq} = 509 kN·m, torsional stiffness $c = 4.7 \cdot 10^6$ N·m, $c = 2.35 \cdot 10^6$ N·m respectively; c – torque M_{torq} = 814 kN·m, $c = 7.52 \cdot 10^6$ N m; 1, 2 – half-coupling cheeks; 3 – pneumatic element; 4 – hub; 5 – shaft.

The second term of the expression (1) describes the static characteristic of the coupling:

Fig. 2 shows the static characteristic of the flexible coupling under study.

$$M(\alpha) = n_{\text{ПБ}} R_{\text{M}} S(\alpha) \times \left[(p_a + p_{uo}) \left(\frac{V_{2o}}{V_{2o} - R_{\text{M}} \int_0^\alpha S(\alpha) d\alpha} \right)^n - p_a \right] \quad (2)$$

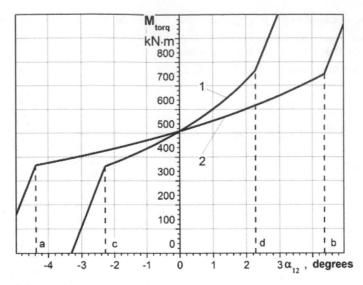

Figure 2. Static characteristic of the flexible rubber-cord shell coupling: 1, 2 - elastic characteristics of couplings, Fig. 1a and Fig. 1b, respectively; ab, cd –coupling movement

The effectiveness of flexible couplings using rubber-cord shells as elastic elements filled with air or water jointly with a hydropneumatic accumulator to limit the shock dynamic loads has been studied. The results are demonstrated on an example of a directly started tumbling mill equipped with a synchronous motor of $N = 4,000\ kW$ power and $n = 75\ rpm$ rotational speed (Fig. 5a).

An equivalent dynamic diagram of the tumbling mill is shown in Fig. 3, where $I_1,\ I_2$ are the rotor moment of inertia and the drum equivalent moment of inertia. M_{em}, M_C are the motor electromagnetic torque and the drum moment of resistance; φ_1, φ_2 are rotation angles; $\alpha_{12} = \varphi_2 - \varphi_1$ is the torsion angle of flexible coupling (the coupling); $M(\alpha_{12})$ is elastic characteristics of the coupling. The pattern of change during the starting period has been presented in (Vinogradov 2004).

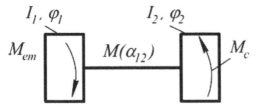

Figure 3. Equivalent dynamic diagram of a synchronous single-motor drive

The elastic characteristics $M(\alpha_{12})$ of the rubber-cord shell couplings are described by the equations (Vinogradov 2004).

Equations for single-motor drive movements are as follows:

$$
\begin{cases}
\dfrac{d\Psi_d}{dt} = U_m \sin\theta - \\[4pt]
\qquad -\left(\dfrac{\Psi_d}{x''_d} - \dfrac{\Psi_f}{x''_{df}} - \dfrac{\Psi_{ed}}{x''_{ded}}\right)r_a - \Psi_q\omega_1; \\[10pt]
\dfrac{d\Psi_q}{dt} = U_m \cos\theta - \left(\dfrac{\Psi_q}{x''_q} - \dfrac{\Psi_{eq}}{x''_{q\ni q}}\right)r_a + \Psi_d\omega_1; \\[10pt]
\dfrac{d\Psi_f}{dt} = U_f - \left(\dfrac{\Psi_f}{x''_f} - \dfrac{\Psi_d}{x''_{df}} - \dfrac{\Psi_{ed}}{x''_{fed}}\right)r_f; \\[10pt]
\dfrac{d\Psi_{ed}}{dt} = -\left(\dfrac{\Psi_{ed}}{x''_{ed}} - \dfrac{\Psi_d}{x''_{ded}} - \dfrac{\Psi_f}{x''_{fed}}\right)r_{ed}; \\[10pt]
\dfrac{d\Psi_{eq}}{dt} = -\left(\dfrac{\Psi_{eq}}{x''_{eq}} - \dfrac{\Psi_q}{x''_{qeq}}\right)r_{eq}; \\[10pt]
\dfrac{d\theta}{dt} = 1 - |\omega_1|;
\end{cases}
$$

$$M_{em} = \Psi_q i_d - \Psi_d i_q; \qquad (3)$$

$$
\begin{cases}
\dfrac{d\omega_1}{dt} = \dfrac{1}{H_1}\left[M_{em} - M(\alpha_{12})\right]; \\[10pt]
\dfrac{d\omega_2}{dt} = \dfrac{1}{H_2}\left[M(\alpha_{12}) - M_C\right]; \\[10pt]
\dfrac{d\varphi_1}{dt} = \omega_1; \qquad \dfrac{d\varphi_2}{dt} = \omega_2,
\end{cases}
$$

111

where Ψ_d, Ψ_q are the magnetic-flux linkage of the stator windings on the longitudinal and transverse axes; Ψ_f is magnetic-flux linkage in field windings; Ψ_{ed}, Ψ_{eq} are magnetic-flux linkage in the damper windings along the longitudinal and transverse axes; U_m is an amplitude of the phase voltage; U_f is field winding voltage; ω_1, ω_2 are angular frequencies of rotation of the motor shaft and drum, respectively; H_1, H_2 are moments of inertia of the motor and the drum; M_{em} is electromagnetic torque; M_C is moment of resistance; θ is an angle between the vector of emf induced in the stator by the field current and the voltage vector ; r_a, r_f, r_{ed}, r_{eq} are active resistances of the stator phase, field winding, damper winding along the longitudinal and transverse axes , respectively; x''_d, x''_f, x''_{ed}, x''_{df}, x''_{ded}, x''_{fed}, x''_q, x''_{eq}, x''_{qeq} are subtransient reactances.

Fig. 4 shows the computed magnitudes of the torque in the drive equipped with a conventional flexible coupling (Fig. 4a) and a coupling equipped with rubber-cord shells as pneumatic elements.

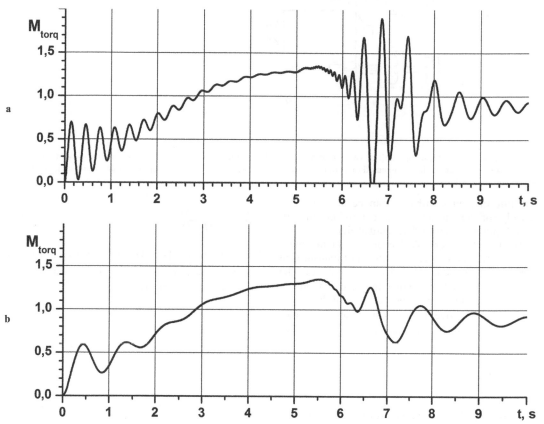

Figure 4. Torque changes in the drive mechanical system during a direct start of the synchronous motor: a, b – with a conventional flexible coupling and a coupling using rubber-cord pneumatic elements, respectively.

Along with the compensation for a significant radial, axial and angular displacements and limiting dynamic impact and resonant loads, the use of rubber-cord shell couplings in heavy machinery drives, where it is necessary to branch the power flow, will allow to solve the problem of load balancing between transmission lines (Vinogradov, Yemelyanenko 2011; Vinogradov 2010).

Fig. 5 shows the diagrams of tumbling mill drives equipped with flexible rubber-cord shell couplings.

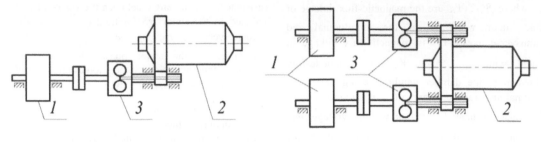

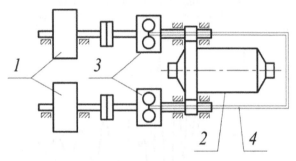

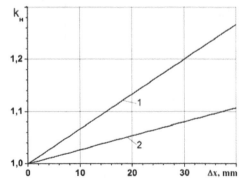

a

b

c

Figure 5. Diagrams of tumbling mill drives with rubber-cord shell couplings: a, b – single-motor and twin-motor drives; c – rubber-cord shell cavities connected to a common conduit. 1 – motor; 2 – drum; 3 – elastic rubber-cord shell coupling; 4 – common conduit

Uniform load distribution can be achieved discretely or automatically via controlling the air (fluid) pressure in the rubber-cord shells.

Using rubber-cord shell couplings in the twin-motor drive can reduce the stiffness of transmission lines and thus reduce an uneven distribution of the load between the transmission lines in each motor.

Uneven load distribution factor will be determined by the value k_H showing how many times the torque reacted by the most loaded motor exceeds the nominal value equal to $0.5M_C$

$$k_H = \frac{M_{max}}{0.5M_C}$$

Assuming that the uneven distribution of the load is caused by the rotor misalignment angle $\Delta\varphi$ we can write

$$k_H = 1 + \frac{c\Delta\varphi}{M_C} \qquad (4)$$

where c is linearized torsional stiffness of the coupling. Expressing $\Delta\varphi$ through a linear magnitude using the rubber-cord shell installation diameter D_1 we can present the equation (4) as

$$k_H = 1 + \frac{c}{M_C}\frac{\Delta x}{D_1} \qquad (5)$$

Fig. 6 shows the relationship between k_H and Δx for a twin-motor drive, equipped with rubber-cord pneumatic elements (Fig. 5b).

Figure 6. Uneven load distribution factor as a function of the linear value of the rotors mismatch along the diameter of the pneumatic elements mounted: 1, 2 – for couplings without and with the additional volume respectively.

113

3 CONCLUSIONS

The analysis of the calculation results suggests that when applied in twin-motor synchronous drives, flexible couplings equipped with rubber-cord pneumatic elements allow evenly distributing the load, reducing dynamic impact and resonant loads. Merely by decreasing the stiffness of mechanical transmissions, the flexible couplings can significantly reduce the uneven distribution of loads.

REFERENCES

Borodulin I.P., Brener E.D., Grechishchev E.S. et al., ed. Panova N.I. 1976. *Locomotives: design, theory and calculation* (in Russian). Moscow: Mashinostroyeniye: 544.

Pastukhov V.A., Rosin G.S. 1977. *Engineering calculation method for balloon rubber-cord shells designed for vibration isolators* (in Russian). Moscow: Nauka: Methods and tools for human vibration protection: 48- 102.

Vinogradov B.V. 2011. *On the dynamic characteristics of viscoelastic systems with rubber-cord shells* (in Russian). Vibrations in engineering and technology, Issue 1 (61): 20- 24.

Vinogradov B.V. 2004. *Dynamics of tumbling mills. Monograph* (in Russian). Dnipropetrovs'k: UDHTU: 127.

Vinogradov B.V., Yemelyanenko V.I. 2011. *Synchronous twin-motor drive in tumbling mills* (in Ukrainian). Patent 102137 Ukraine, MPK7 B02C 17/24 Applicant and patentee holder: SHEI "Ukrainian State University of Chemical Technology"- No.a201109919; application date 10.08.2011; published 10.05.2012, bulletin № 9. – 3 p.: Il.2

Vinogradov B.V. 2010. *Synchronous twin-motor drive in tumbling mills* (in Ukrainian). Pat. 96521 Ukraine, MPK7 B02C 17/24 Applicant and patentee holder: SHEI "Ukrainian State University of Chemical Technology"- No. a201007858; application date 23.06.2010, published 10.11.2011, bulletin № 21, 2011. – 2p.: Il.1

Power Engineering, Control and Information Technologies in Geotechnical Systems – Pivnyak, Beshta
& Alekseyev (eds)
© 2015 Taylor & Francis Group, London, ISBN 978-1-138-02804-3

Determination of optimal configuration
of effective energy supply system with multiple sources

Yu. Khatskevych
State Higher Educational Institution "National Mining University", Dnipropetrovs'k, Ukraine

ABSTRACT: For today's world the solution of problems of energy saving, energy efficiency has become a matter of vital importance. One way of solving this problem is the use of different energy sources including alternative and renewable. It is very important to ensure the effectiveness of their operation. The core elements in the process of selection of multiple sources to improve energy efficiency of energy supply system are analyzed. This paper presents an approach to determine the optimal energy supply mix including centralized and decentralized energy sources for large-scale geotechnical and industrial objects. It is proposed an approach to select types and power of the sources in the energy system that provide energy effectiveness based on evolutionary algorithm of random search. Functions of generation and selection of solutions are formulated. The proposed algorithm provides a high degree of flexibility and could be used as well for heating and electricity supply systems to determine the optimal configuration. An example of application of proposed algorithm for an existing industrial enterprise is given.

1 INTRODUCTION

To enhance reliability of energy supply and increase technical and economical characteristics of operation it is advisable to apply some energy sources for heat (cooling) and electrical energy. This approach is especially common for renewable sources. Joint work of the centralized and decentralized energy sources is another technical decision. It is necessary to take into account that in this case the mode of operation and power consumption of one energy source influence the characteristics of the whole energy system. Therefore, for effective operation of energy system with some sources of electrical and heat energy it is necessary to work out methods of making decisions and control algorithms for the energy system as a whole.

Nowadays the issues of autonomous energy supply (heat and electrical energy) are quite completely studied concerning cottages and small buildings (Gruber, Fernández & Prodanovic 2013). Possibilities of using of different energy sources both traditional and alternative and various equipment and systems of intelligent power management are investigated. However, a standard methodology for selecting the most preferable energy sources from the variety of equipment as well as control algorithms depending on energy needs for such complex systems are not thoroughly worked out.

It is important to take into account that with increasing power of the object increases the complexity of the problem of the selection of sources for it. For complex objects such as geotechnical systems consisting of dozens of buildings, structures and installations, it is necessary to develop special approaches for selection of optimal configuration of energy supply system.

2 PROBLEM DEFINITION

Core elements in the problem of energy sources selection could be distinguished.

1. Collection of data about actual seasonal and daily energy loads.

Collection should be implemented for some climate and geographical zones that will be chosen as typical for solutions realization.

2. Composition of a database for key characteristics of equipment for generation, distribution and transformation of energy taking into consideration probable operation modes.

It is necessary to compose database using manufacturers' data for equipment that could be used for generation of electrical and heat energy, heat recuperation and energy storage, systems of energy transformation and distribution. The data should include key energy, technical and cost characteristics for each type of the equipment.

3. Selection of the group of most commonly used effective power equipment for further research their energy and technical characteristics.

Justification of the selection of most commonly used models and types of equipment for detailed study of characteristics, analyze of joint work in energy system and so on. Working out of approach to estimate accuracy of results when other types or models of equipment are applied.

4. Classification of selected equipment according to conditions of effective usage depending on the key factors: climate, natural resources, tariffs, legal aspects of power energetic in region and so on.

Composition of a database of economical and legal aspects of power energetic in region. Determination of a set of equipment that is possible to use for energy supply in residential buildings according to their purpose, settlement, climate conditions, etc.

5. Modeling of storage and usage processes of generated energy.

Mathematical and physical modeling of the storage and usage processes of energy that was generated by own sources. As possibilities to store energy heat accumulating in water tanks or in soil and hydrogen accumulating with use of fuel cells could be studied.

6. Modeling of work of different combinations of equipment taking into account specific operation conditions: tariffs, climate, modes of energy consumption, construction characteristics of the building, availability and features of centralized energy system, etc.

Mathematical and physical modeling of the modes of operation of energy supply systems for residential buildings that could contain some energy sources, with the possibility to store and recuperate energy or change the mode of energy consumption. As a result technical, economical and energy characteristics of energy supply system including equipment for transportation, distribution and transformation of energy should be received for various specific operation conditions.

7. Selection of a criterion of energy or economy efficiency and making a decision about the most expedient type of energy supply, mode of operation and control for sources.

Justification of the criteria of energy efficiency of building modernization taking into account energy and economy characteristics of equipment. Working out of a decision making algorithm for selection the most expedient energy supply equipment and modes of its operation.

8. Coordination of work of autonomous and centralized energy sources.

Depending on tariffs and legal aspects of power energy operation in a specific region there is a need to develop algorithm of coordinated work of autonomous and centralized energy sources as a joint combined energy system. It is necessary to take into account possibilities of energy storage and influence on quality coefficients of such systems.

In this work a part of these complex methodic will be considered, namely a justification and selection of heating energy sources type and power that allow to minimize energy recourses consumption.

3 THE MAIN CONTENT OF THE WORK

Solving of the problem of selection of the composition and power of energy system elements began with the simplest option – a binary system, for which the technical and economic characteristics of the different modes were examined.

During joint research work with the Esslingen University of Applied Sciences (Germany) an operation of binary heating systems, consisting of a heat pump and boiler were analyzed. It were studied boilers with different principles of work using coal, pellets, gas and electrical boilers. To determine the optimum ratio of capacity of the heat pump and boiler were built dependences on capital and operating costs of binary system per 1 kW of power. Capital costs were determined by analytical dependences obtained by approximation of data from equipment manufacturers (Khatskevych, Koshelenko 2013). It was considered that the heat pump in the binary system is the main source of heat to cover the heat load for outside temperatures above -15 ° C. If the heat pump capacity is not enough to cover the load boiler is switched on.

Our studies show that dependences on consumption and capital costs on the ratio of the power of different energy sources do not have explicit minimum. Thus, determination of the optimum ratio of sources capacity in binary system on these indicators is inappropriate. Change of the value of a particular type of the fuel or equipment can make a decision taken by such dependencies wrong. Therefore it is proposed to develop an approach to determine the optimal power of the elements of the heating system with multiple sources.

In general, the problem of selection the optimal structure and power of energy supply system are nonlinear optimization task with many discrete unknowns. To improve the accuracy and adequacy it is proposed to work out a model for decision making based on an algorithm of random search.

4 STATEMENT OF THE PROBLEM OF SELECTION OF THE ENERGY EFFICIENT ENERGY SUPPLY SYSTEMS

Proposed the following formulation of the problem of selection of the sources of the energy supply system. For a given object heating demand Q is known. The types of energy sources that could be used at this facility are selected according to the technological possibilities and availability of resources. The number of systems that may be applied is denoted n.

For each type of power source a sequence of discrete data of equipment power is given: $q_{1j}, q_{2j}, ..., q_{mj}$, where $m(j)$ - the quantity of types and sizes of equipment of power system with number j ($j = 1, ..., n$). For this set of equipment are known:

• capital costs of implementation $K_{1j}, K_{2j}, ..., K_{mj}$;
• quantities of fuel and energy resources for operation of equipment $\Pi_{1j}, \Pi_{2j}, ..., \Pi_{mj}$, these values have to be calculated taking into account under specific temperature of the region;
• operating costs of the equipment $E_{1j}, E_{2j}, ..., E_{mj}$.

It is necessary to determine the capacity of each type of energy source that will be used in energy system of the object. It is proposed to present them in the form

$$x^1, x^2, ..., x^i, ..., x^n, \qquad (1)$$

where

$$x^i \in \left\{ q_{1j}, q_{2j}, ..., q_{mj} \right\}. \qquad (2)$$

Desired values are limited:

$$\sum_{i=1}^{n} x^i = Q, \qquad (3)$$

In the solving of the problem of equipment selection it is necessary to take into account heat load for hot water supply Q_{hw}, heating of air in a ventilation system Q_v, etc. Then the desired value of power energy have to satisfy to a requirement:

$$\sum_{i=1}^{n} x^i = Q + Q_v + Q_{hw}, \qquad i=1,2,...,n. \qquad (4)$$

To create an energy saving system it is necessary to provide a minimum of energy costs in its operation:

$$\Pi = \sum_{i=1}^{n} \sum_{j=1}^{m} \Pi_{ij} \rightarrow \min. \qquad (5)$$

Another solution would have the task while minimizing operating costs and requirement for it would be:

$$E = \sum_{i=1}^{n} \sum_{j=1}^{m} E_{ij} \rightarrow \min. \qquad (6)$$

Type of the objective function is determined by a responsible person.

It is necessary to take into account the limitations that usually exist for capital costs of the system of energy supply:

$$K = \sum_{i=1}^{n} \sum_{j=1}^{m} K_{ij} \leq K_{\max}, \qquad (7)$$

K_{max} – the highest allowable value of capital costs.

To solve this problem it is proposed to use the random search method – namely evolutionary. This method is useful for solving tasks of the nonlinear optimization.

In this case, the search is to be kept in the space of possible elements of heating supply system (1). It is necessary to find minimum of function (5) or (6) with limitations (4) and (7).

For electrical system statement of the problem is the same. It is proposed to make search for sources of the electrical system separately from the heating system. It will allow to use the same algorithm. For electricity demand Q^e the number of systems that may be applied is denoted n^e. Optimal configuration will consist of possible elements of power supply system (1) that provide minimum of function (5) or (6) with limitations (4) and (7) that are calculated for equipment of the power supply system.

5 ALGORITHM OF THE SELECTION OF ENERGY EFFICIENT HEATING SYSTEM

It is proposed an evolutionary search algorithm in the form:

$$X_{sk} = S(G(X_{sk-1})), \quad s = 1,...,N_b, \quad k = 1,2,... \qquad (8)$$

where X_{sk} – set of the best solutions that have been selected on the k-th step of the s-th branch of the iterative search process; S (.) – function of decisions selection; $G\,(.)$ – function of solutions generation; s – serial number of branches of the evolutionary search; k – serial number of the step of iterative search process; N_b – total number of branches of evolution.

At each step of the evolutionary search the best solutions at the point of binary relation are selected. In the task of minimizing the use of energy resources with limitations on capital costs binary relation R_S of selection looks:

$$x^1 R_s x^2 = \left[K(x^1) \le K_{\partial on} \cap K(x^2) > K_{\partial on} \right] \cup$$
$$\left[K(x^1) \le K_{\partial on} \cap K(x^2) \le K_{\partial on} \cap K(x^1) < K(x^2) \right] \cup$$
$$\cup \left[K(x^1) \le K_{\partial on} \cap K(x^2) \le K_{\partial on} \cap E(x^1) < E(x^2) \right]. \tag{9}$$

Parameters of evolutionary search control have been formulated. Solutions that are generated for each desired variable in the branch of the evolutionary algorithm are as follows:

$$x_a = \left\{ x_a^1, x_a^2, \dots x_a^n \right\}, \ a = \overline{(N_\kappa + 1), N_b}, \tag{10}$$

where a - number of a solution that was selected from the best of the previous iteration step; N_k – the number of the best decisions are selected at each step of the search.

Usually there is a set of possible power equipment capacities and sizes. Therefore, the desired solution of capacity of the heating elements should be considered as discrete variables. It is proposed to organize generation of the values in the random search as follows:

$$x_a^i = x_r^i (1 + \zeta^i), \tag{11}$$

where r – random number that is chosen with equal probability from the set of the best decisions of the previous iteration step $r \in \{1, 2, \dots, N_\kappa\}$; ζ^i – random variable that has a normal distribution with zero expectation and variance

$$\sigma_i^2 = \frac{1}{N_\kappa N_b - 1} \sum_{a=1}^{N_\kappa} \sum_{i=1}^{N_b} \left(x_{la}^i - x_0^i \right)^2, \tag{12}$$

where x_{la}^i – the best solutions selected in the previous step of the iteration process; x_0^i – mean values of variables selected in the previous step.

$$x_{0i}^i = \frac{1}{N_\kappa N_b} \sum_{a=1}^{N_\kappa} \sum_{i=1}^{N_b} \left(x_{la}^i \right)^2. \tag{13}$$

The values of the variables obtained by the expression (11) are rounded off to the nearest value of the set of possible equipment power. Thus,

$$x_a^i = q_j^i, \tag{14}$$

for which

$$\Delta = \left| x_a^i - q_j^i \right| = \Delta_{\min}^i. \tag{15}$$

It should be noted that the proposed approach can be used for complex objects with many buildings and energy sources. Their energy costs can be considered together in order to reduce energy consumption in a separate building, and in general for complex object. In this case the representation of the desired variables in the algorithm (power of the elements of heating and electricity supply systems) will be changed.

For the proposed search algorithm a program of calculation is worked out.

6 AN EXAMPLE OF IMPLEMENTATION OF THE PROPOSED ALGORITHM

It is considered on geotechnical object consisting of workshops and administrative buildings. The total area of the enterprises is about 14,000 m^2.

The heating load of the enterprise is 28 MW, of which for heating are 21 MW and for air in ventilation system are 7 MW.

The existing heating system of the enterprise is centralized, from a boiler running on natural gas.

Annual consumption of gas for heating and ventilation of 6.9 million m^3.

There are considered the following options of the technically feasible solutions for modernization of the heating system:

1 centralized heating system from the boiler of the enterprise;

2 decentralized heating system with mini-boilers on gas fuel;

3 decentralized heating system with mini-boilers on electricity;

4 decentralized heating system with mini-boilers operating on liquid fuel (for example, diesel);

5 low-temperature infrared heating system that runs on gas fuel;

6 high-temperature infrared heating operating on electricity;

7 system using waste heat energy of industrial processes and buildings.

For each type of the heating source was collected information about possible variants of equipment, its power, capital and operating costs. The goal was to minimize operating costs that are mostly consisting of fuel costs.

Optimal configuration of the heating system was represented as (1), provides minimum of the function (5) with the (4) and (7). Generation and selection of the solutions were made according to equations (8 – 15).

It is made the following technical solution about optimal configuration of the heating system: heating of five of the workshops it is advisable to carry out with the help of low-temperature infrared heaters; 2 workshops are heated through their own waste heat, one workshop is heated by infrared heaters and waste heat; for part of the administrative offices it is advisable to build a mini-boiler operating on gas fuel; for certain administrative buildings it is expedient to leave the existing centralized heating system from the boiler plant. Plan of the enterprises with recommended types of heating system for each building according to its heating demand are shown on Fig. 1.

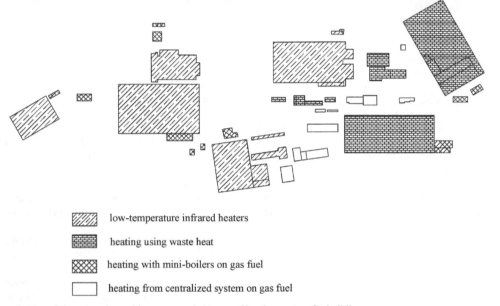

▨	low-temperature infrared heaters
▦	heating using waste heat
▧	heating with mini-boilers on gas fuel
☐	heating from centralized system on gas fuel

Figure 1. Plan of the enterprises with recommended types of heating system for buildings

Due to proposed decision reduction of natural gas consumption is 2.7 million m^3 per year. The payback period of the project is 1.8 years.

7 RESULTS

The analysis of possibilities of selection of the sources for energy supply systems for complex objects such as geotechnical systems or large-scale industrial enterprises allow to conclude that the special algorithms for this purpose should be worked out. It is proposed the algorithm to select types and power of the sources in the energy system that provide energy effectiveness based on evolutionary algorithm of random search. For this algorithm statement of the problem, rules of solution generation and selection are formulated.

The results of application of the proposed algorithm for selection of the sources of heating system for large-scale industrial enterprise are represented. Energy effective heating system that was chosen allows to decrease energy consumption of the enterprise on 40 % with the payback period of 1.8 years.

The proposed algorithm provides a high degree of flexibility and could be used as well for heating and electricity supply systems to determine the optimal configuration.

REFERENCES

Gruber J., Fernández J. & Prodanovic M. 2013. *Supply Mix Optimization for Decentralized Energy Systems.* Open Journal of Applied Sciences, Issue 3, Vol. 2B: 5-11.

Khatskevych Yu.V., Koshelenko Ye.V. 2013. *Study of duration of heat load on selection of equipment for bivalent heating systems* (in Ukrainian). Scientific reports collection of the conference "Perspectywy rozwoju badan naukowyh w 21 wieku" (Szczecin, 27.02.2013 - 28.02.2013): 23-28.

Power Engineering, Control and Information Technologies in Geotechnical Systems – Pivnyak, Beshta
& Alekseyev (eds)
© 2015 Taylor & Francis Group, London, ISBN 978-1-138-02804-3

Simulation of the energy-effective operating modes of the mine main pumping

N. Rukhlova
State Higher Educational Institution "National Mining University", Dnipropetrovs'k, Ukraine

ABSTRACT: Actuality of the use of main dewatering plant of coal mine as an effective controllable load is shown. The simulation algorithm of the pumps operating modes of the mine pumping which allows to increase efficiency of regulation of the electro-consumption modes with its help is developed.

1 INTRODUCTION

The mine pumping is a classic controllable load, however efficiency of its use in this quality remains enough high not for all conditions. Participation of the main dewatering plant (MDP) at regulation of the electro-consumption modes must be economic advantageous to the both electrical customer (to the mine) and producer (to the power grid) on condition of the use of the differential rate on electric power. On this basis, an economic effect from the change of the pumping operating modes with the aim of pumps disconnection in the hours of the peak demand and intensive pumping-out of water in the nightly hours must be high (Razumniy, Rukhlov & Rukhlova 2013).

2 AIM OF THE WORK

Aim of the work – development of the simulation algorithm of the energy-effective operating modes of pumps of the main dewatering plant of coal mine.

3 ACCOUNT OF BASIC MATERIAL

Functioning efficiency of the mine pumping is determined by the its operating mode, which relies on such descriptions, as a mine water inflow, lodgment capacity, parameters of the basic MDP's equipment, and also by the control by this mode. Energy efficiency of the water pumping-out process increases at regulation of the electro-consumption modes of dewatering plant, which is possible only for the known operating modes of pumps. Such modes can be got by the simulation in accordance with algorithm the simplified flow-chart of which is resulted on the Fig. 1 (Razumniy, Rukhlov & Rukhlova 2014). The developed algorithm allows to get the most energy-effective functioning mode of dewatering plant, taking into account the actual parameters of its equipment, and also realize it by the maintenance staff.

In the block 1 initial information for the concrete mine conditions is entered: pumping level, total volume of lodgment, quantity and diameter of piping, quantity of pumps and their nominal parameters, and also the quantity of pumps switching on is limited.

In the block 2 the useful volume of lodgment taking into account silting is estimated, and also the indexes of pressure characteristic of piping system and the pumping unit parameters are determined.

In the block 3 the equivalent resistance of piping system for the standard circular scheme is determined with possibility of specification of the actual inner diameter of pipeline.

In the block 4 the pumps operating parameters depending on the their functioning mode are determined. Also its re-calculation taking into account the actual technical condition of equipment is possible. The coefficient of efficiency is the basic parameter characterizing high-quality functioning of pump, which relies both on the operating mode of pumping unit and its technical condition.

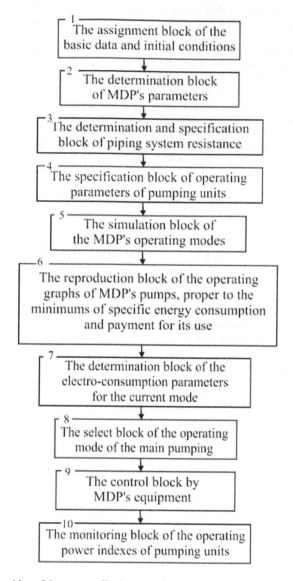

Figure 1. The simulation algorithm of the energy-effective operating modes of main dewatering plant

In the block 5 the operating modes of pumping units taking into account the technological conditions of concrete object are simulated.

In the block 6 the simulated variants are sort out and selected from them the cyclic operating modes of pumps for the set time period (usually day's interval). Among the got great number of the graphs select the operating mode of dewatering plant, which corresponds to the criterion of minimum specific energy consumption on water pumping-out at the minimum value of payment for its use.

Expediently to reproduce the pumping operating modes, the values of the given criteria for which differ from the minimum value on the definite percent. That percent can be set to discretion of the maintenance staff (as a rule 1–5 %). Such approach extends the select possibility of the proper operating mode of pumps for the concrete conditions, as a value of minimum criterion of specific energy consumption not always corresponds to the lowest payment for its use.

In block 7 for every current operating mode of pumping the values of general and specific energy consumption, payment for consumable electric power, volume of pumped out water and others parameters are accounted.

In block 8 from the simulated graphs within the limits of the set percent restriction from the minimum value of specific energy consumption (block 6) the pumping operating mode is selected.

In the block 9 information got in the block 8 will be realized. The simulated operating modes of pumps and their parameters are sent to the maintenance staff (to the motorman of pumping units, mechanic and others) which will realize the selected operating mode by hand or automatically, at presence of the proper control system.

In the block 10 basic power parameter of MDP operating – specific energy consumption on water pumping-out is monitored. The monitoring of its numeral values and comparison with previous (typical) allows to do the conclusion in relation to the reason of change of specific consumption (reduction of pipeline inner diameter as a result of mineral sediment on its walls, worsening of technical condition of pumping units etc.). The conclusion and decision concerning the subsequent actions are accepted by the maintenance staff.

In accordance with algorithm for the cited conditions and limitations the quantity of possible variants of pumping work is determined. There is a possibility of construction of the MDP's work graphs with the cyclic operating mode of pumps for the calculated period of time. Is also reproduced to any from the possible graphs of lodgment filling and proper to it the electrical loading diagram of MDP. As a result the mode with the minimum specific energy consumption on condition of the lowest payment for used electric power is selected. Except for this the quantity of the simulated variants proper to the given criteria accounts, which is determined by the size of percent limitation (tolerance) from their minimum values. For the selected (current) mode the electro-consumption parameters are determined: payment for used electric power for day, general and specific energy consumption for the graph and average weighted electro-consumption by the pump characteristics.

As an example on the Fig. 2 the diagram of the possible MDP's operating modes is resulted. We can see a great number of the pump operating modes of pumping at the identical initial conditions (quantity of pumps $N_p = 2$ and quantity of its switching on for day's interval is limited about 6). At the water volume in lodgment $V_0 = 1200$ m^3 at the day beginning $t_0 = 0$ hour is offered three variants of pumps operating: 0 – none pump works, 1 – one works, 2 – two pumps work. Also the change of the operating mode of pumping units is possible every hour (t_{ch}). Thus the simulation of the operating modes of pumping units is realized in the set limits of lodgment parameters and with the set requirements in relation to disconnection of pumps in the periods of the maximal loading in power grid (in the peak area P).

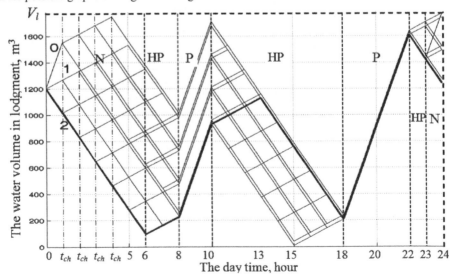

Figure 2. The diagram of the possible operating modes of the main dewatering plant with two pumps

In the general case the area of all possible operating modes of MDP is limited by rectangle parties of which are: on the vertical line – axis of water volume in lodgment V (moment of initial time) and line of final time, during which the operating modes are simulated (expediently to accept day's interval); on the horizontal line – axis of time, below than which the pumps runs free, and line, limiting a lodgment volume V_l possible for filling, higher than which it will be overfilled.

The fat broken line (Fig. 2 and 3) as an example select one of the possible cyclic functioning modes of dewatering plant on day's interval. In this case in the nightly period N from 00^{00} to 06^{00} hour and from 23^{00} to 24^{00} hour – two pumps work; in the period of half-peak loading HP from 06^{00} to 08^{00} and from 10^{00} to 13^{00} hour – one pump works, and from 13^{00} to 18^{00} and from 22^{00} to 23^{00} hour – two pumps; in the period of the maximal (peak) loading P from 08^{00} to 10^{00} and from 18^{00} to 22^{00} hour – the pumps do not work.

Developed by the simulation method the algorithm of the MDP functioning in the controllable load mode allows to get the great number of the possible operating modes of pumping at the set conditions and limitations, and also to select from this great number most energy-effective. Within the framework of algorithm it is possible to set any interval of change of the operating mode of the MDP's pumps for a day long, that enables to select the mode at which in the period of maximal loading in power grid the pumps will be disconnected. In this case the day's energy consumption and explicit costs on its payment will be minimum. In addition, expediently to limit a general quantity of pumps switching on for a day long, that will allow to prolong a operation period of pumps and their drive engines.

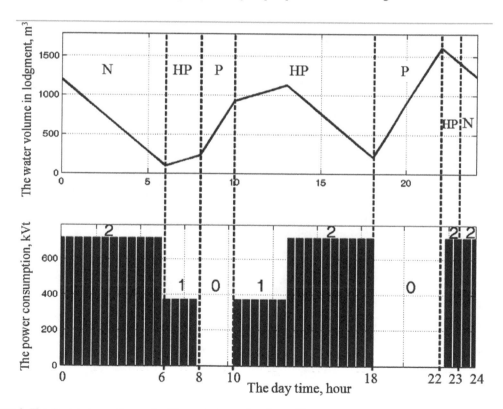

Figure 3. The simulated operating mode of the main dewatering plant with two pumps

4 CONCLUSIONS

The developed algorithm of functioning of the main pumping allows to simulate its energy-effective operating mode for any technological characteristics of dewatering plant and technical parameters of pumps. Such mode is cyclic for day's interval and is characterized by the minimum specific energy consumption on water pumping-out at condition of the payment decrease for its use.

REFERENCES

Razumniy Yu.T., Rukhlov A.V., Rukhlova N.Yu. 2013. *Improvement of energy efficiency of the main dewatering plant of a coal mine* (in Russian). Dnipropetrovs'k: Scientific bulletin of National Mining University, Issue 5: 67–71.

Razumniy Yu.T., Rukhlov A.V., Rukhlova N.Yu. 2014. *The control method of the main dewatering plant of a coal mine* (in Ukrainian). Patent 93990 Ukraine, MKY F04D 27/00, applicant and patent holder: State Higher Educational Institution "National Mining University", declaration date 30.04.2014, published 27.10.2014, bulletin № 20.

A new method to solve a continuous facility location problem under uncertainty

S. Us
State Higher Educational Institution "National Mining University", Dnipropetrovs'k, Ukraine

ABSTRACT: The optimal location problem for enterprises in a given region has been studied under uncertainty. New mathematical models with multi-valued cost functions have been formed. Approaches to solve the problem of optimal placement of enterprises and find their service areas have been proposed. A new solution algorithm based on a continuous set partitioning method has been suggested for a location problem with multi-valued cost function. The results of computer experiments have been presented.

1 INTRODUCTION

A great number of papers in the field of operation research are dedicated to solving the problem of planning and object position. Such tasks have widespread practice application since the area of placement can have different structure and the term «object» may be interpreted in a wide sense (Farahani & Hekmatfar 2009; Drezner & Hamacher 2001). The examples of such tasks are the location of various service points (hospitals, shops, fire stations, enterprises of various types, etc.), developing enterprise master plans, engineering PC-boards, constructing aircrafts, irrigation problems, designing the networks of mobile communications etc (Kiseleva & Shor 2005; Us 2013).

All location problems can be divided into two big classes: the tasks of interrelated object placement and the tasks of placement-allocation (the tasks of enterprise placement).

The first class includes the tasks that have preliminarily specified connection structure between objects, e.g Veber problem. The tasks of the second-class have no connection between placed objects –"suppliers", and distribution of fixed objects-"clients" is carried out. The examples of such tasks are the following: p-median and p-center tasks, the simplest placement task etc. Various classes of placement-distribution tasks are generated by different structure of placement area and assumptions about possible facility location and objects-clients. Enterprise location problems in case of a discrete set of possible facility locations have been investigated for more than a century, but developing efficient algorithms to solve these tasks is still a topical task. The tasks with displaced objects to be located in any point of a specific area are more complicated. The distance metric between the objects is defined and the clients continuously provide the filling of some areas.

The research into these tasks was carried out and reported in literature (Kiseleva & Shor 2005; Corley et al 1971).

Let us formulate the task of optimal partitioning of continuous set in the following way .

Supposing, there is a set of consumers of some homogeneous product allocated in the area Ω. The finite number N of producers placed in the isolated points τ_i, $i = \overline{1, N}$ of the area Ω forms the system of points τ_1, τ_2, ..., τ_N ; besides, the coordinates of some points or all of them can be earlier unknown. The demand $\rho(x)$ for the product at each point x of the area Ω and product delivery cost $c_i(x, \tau_i)$,

$i = \overline{1, N}$ from a producer τ_i to a client x are known. Let us assume that the producer's profit is only dependable on transportation costs. The capacity of the producer i is defined by a total demand of the service clients and should not be higher than specified volumes b_i, $i = \overline{1, N}$. The area Ω is to be subdivided into service zones by each of the producers, i.e. into sets Ω_1, Ω_2, ..., Ω_N, to minimize the total costs spent on the product delivery.

Mathematical model of the stated problem can be presented as follows (Kiseleva & Shor 2005):

Let Ω be a bounded Lebesgue measurable set in an n-dimensional Euclidean space E^n. It is required that the set be divided into N Lebesgue measurable subsets $\Omega_1, \Omega_2, ..., \Omega_N$, and subset centers $\tau_1, \tau_2, ..., \tau_N$ are located in the area Ω to let the functional

$$F\left(\Omega_1,\Omega_2,...,\Omega_N,\tau_1,\tau_2,...,\tau_N\right)=\sum_{i=1}^{N}\int_{\Omega_i}c_i\left(x,\tau_i\right)\rho(x)dx \quad (1)$$

reach the maximum value under constraints

$$\int_{\Omega_i}\rho(x)dx\le b_i,\quad i=\overline{1,N} \qquad (2)$$

$$mes\left(\Omega_i\cap\Omega_j\right)=0,\ i\ne j,\ i,j=\overline{1,N}, \qquad (3)$$

$$\bigcup_{i=1}^{N}\Omega_i=\Omega. \qquad (4)$$

While investigating such tasks, a set of directions has been formed stipulated by different applications and types of mathematical statements. The following tasks could be referred to: determinated linear and non linear tasks, single- and multi-product tasks in terms of constrains having specified centre location and defining their optimal location option; dynamic tasks with optimality criterion dependable on phase orbits and management of some controlled system; tasks under the conditions of uncertainty where the tools of stochastic infinite-dimensional programming (provided that the part of information has a probabilistic character) or the fuzzy set theory are suggested to remove the uncertainty.

Let us consider placement tasks in terms of fuzzy data. They can be divided into two classes. The tasks with fuzzy obtained partitioning will be referred to tasks of optimal fuzzy partitioning; given data are clearly determined. Such tasks were considered in the works (Kiseleva & Shor 2005; Us, Sadovnikova 1999). The tasks with fuzzy given data are referred to the second class. The decision derived in this regard must be explicit. The papers of such scholars as (Kiseleva, Lebid' 2010; Us 2010, 2011, 2012) are dedicated to investigating these tasks in cases with fuzzy constraints. The problems with a fuzzy objective functional, precisely, cost functions, prescribed inaccurately, are considered in this paper. The interest to such models can be explained by the fact that while solving real problems, it is impossible to prescribe exact delivery cost because of its dependence on many factors; most of them have probabilistic or fuzzy character and cannot be closely determined. Weather and road conditions, technical state of transport, staff qualification; economic conditions etc. can be named as the examples of such factors. In addition, it is difficult to describe functional dependence of delivery cost on these factors decides selecting a singular fixed value makes the model rough. But in most cases it is possible to specify some set of potential cost functions corresponding to the external environment conditions. Mathematical models and methods to solve continuous tasks of optimal placement-distribution under inaccurately given cost functions are proposed in this paper.

2 MATHEMATICAL FORMULATION OF THE PROBLEM AND DESCRIPTION OF THE SOLUTION METHOD

2.1 Continuous placement problem with multi-valued cost function

Supposing, there is a set Ω of consumers of some homogeneous product. They are placed in area Ω continuously. The finite number N of producers placed in isolated points τ_i, $i=\overline{1,N}$ of the area Ω form the system of points $\tau_1,\tau_2,...,\tau_N$, moreover, the coordinates of some of these points or all of them can be earlier unknown. The demand $\rho(x)$ for the product at each point x of the area Ω is known and the set of feasible values of delivery cost function $c_i\left(x,\tau_i\right)=\left\{c_i^1\left(x,\tau_i\right);\ c_i^2\left(x,\tau_i\right);...c_i^{M_i}\left(x,\tau_i\right)\right\}$, $i=\overline{1,N}$ from a producer τ_i to a client x is known as well. The producer profit is assumed to be dependable on transportation costs alone. The capacity of the producer i is defined by a total demand of service clients and must not be higher than given values b_i, $i=\overline{1,N}$. Area Ω is to be divided into service zones of each producer, i.e. into subsets Ω_1, Ω_2, Ω_N to minimize the total costs on the product delivery.

The following mathematical model corresponds to a set task.

Let Ω be a bounded Lebesgue measurable set in an n-dimensional Euclidean space E^n. It is required that the set be divided into N Lebesgue measurable subsets $\Omega_1,\Omega_2,...,\Omega_N$, and subset centers $\tau_1,\tau_2,...,\tau_N$ are located in the area Ω to let the functional

$$F\left(\Omega_1,\Omega_2,...,\Omega_N,\tau_1,\tau_2,...,\tau_N\right)=\sum_{i=1}^{N}\int_{\Omega_i}c_i\left(x,\tau_i\right)\rho(x)dx \quad (5)$$

reach the maximum value under constraints

$$\int_{\Omega_i}\rho(x)dx\le b_i,\quad i=\overline{1,N} \qquad (6)$$

$$mes\left(\Omega_i\cap\Omega_j\right)=0,\ i\ne j,\ i,j=\overline{1,N}, \qquad (7)$$

$$\bigcup_{i=1}^{N}\Omega_i=\Omega. \qquad (8)$$

Here $c_i(x, \tau_i)$, $i = \overline{1, N}$ is a multi-valued cost function with a set of values as $c_i(x, \tau_i) = \{c_i^1(x, \tau_i); c_i^2(x, \tau_i); \dots c_i^{M_i}(x, \tau_i)\}$; however, before making a decision, it is not clear what exact value the function will be; $\rho(x)$ is a real and non-negative integrable function defined on Ω; b_i, $i = \overline{1, N}$ are given non-negative real numbers, satisfying the condition of the task solubility:

$$\sum_{i=1}^{N} b_i \geq S \ , \ S = \int_\Omega \rho(x) dx \ . \tag{9}$$

$$P(\Omega, N) = \left\{ (\Omega_1, \Omega_2, \dots, \Omega_N, \tau_1, \tau_2, \dots, \tau_N) \left| \begin{array}{l} mes(\Omega_i \cap \Omega_j) = 0, \ i \neq j, \ i, j = \overline{1, N}; \\ \bigcup_{i=1}^{N} \Omega_i = \Omega; \ \int_{\Omega_i} \rho(x) dx \leq b_i, i = \overline{1, N} \end{array} \right. \right\}, \tag{10}$$

$R = (\Omega_1, \Omega_2, \dots, \Omega_N, \tau_1, \tau_2, \dots, \tau_N)$ is some possible partitioning of the set Ω.

Taking into account the situation, it is possible to consider the following as task solutions:

$$P*(\Omega) = \left\{ R(\Omega) \left| \begin{array}{l} \exists R', \ F(R(\Omega)) \leq F(R'(\Omega)) \ \text{для всех} \ c_i^j(x, \tau_i) \in C, \\ \text{и} \ F(R(\Omega)) < F(R'(\Omega)) \ \text{для} \ c_i^l(x, \tau_i) \in C \end{array} \right. \right\}. \tag{11}$$

Such decision arises from interpreting the task $(5) - (8)$ as multicriteria optimization and corresponds to the set of Pareto-optimal decisions.

2. Partitioning of the set Ω into subsets gives an optimal value to the functional (1) in some sense;

Optimality will be understood as reaching one of effective multicriteria task solutions and (or) a functional reaching some value that should satisfy specific requirements and not worse from other decisions in terms of the preferences induced by the task conditions and requirements of people making decisions.

3. Some fuzzy partitioning of the set Ω into subsets (the definition of fuzzy partitioning is given in the works (Kiseleva, Lebid' 2010).

The approaches to provide solution of the task of class 2 will be considered in this paper. Solving the task of optimal partitioning with multi-valued cost function will be considered as partitioning to give optimal (in the stated above sense) value to objective functional (1).

Considering that cost function $c_i(x, \tau_i)$, $i = \overline{1, N}$ for each fixed x and τ_i is some collection of values, the derived task can be interpreted as the task of multicriteria optimization where the majority of criteria is formed as a functional aggregate

2.1 Description of the solution method

The derived task is the problem of infinite-dimensional programming with multi-valued cost function and there are a few possible approaches to be applied to its solution. Obviously, it is useless to consider minimizing in this task as an ordinary notion, because the functional value is some set of numbers rather than a number. Thus, it should stipulate what task solving means.

According to (Kiseleva & Shor 2005) $P(\Omega, N)$ is denoted as a set of all possible task partitionings $(5) - (8)$:

1. The aggregate $P^*(\Omega)$ of possible partitionings of the set Ω satisfying the conditions:

derived by setting one of the possible collections of cost functions into initial functional. The next interpretation is the task of selection (assume that only one representative of this set is an actual one to be chosen among the whole set of cost functions).

One of possible approaches applied to solution of such tasks is converting the given task with multi-valued objective function into a simple task of mathematical programming by transforming a given set of objective functions into a single compromise objective function.

The simplest method of such transformation is choosing one representative from a function set. It is obvious, that the methods of choosing this representative are different depending on the available input information.

First of all, it should be noted that under the conditions of complete uncertainty, when a person making a decision has no information regarding possible preferences in terms of cost set, any of these values can be chosen, for example:

− minimum:

$$c_i(x, \tau_i) = \min_{\theta_j} c_i^j(x, \tau_i), \ i = \overline{1, N}, \ x \in \Omega. \tag{12}$$

Such choice corresponds to the optimistic position (forecast), because computing includes the minimum cost.

– maximum:

$$c_i\left(x,\tau_i\right) = \max_{\theta_j} c_i^j\left(x,\tau_i\right), \qquad (13)$$

Such choice corresponds to the pessimistic position (forecast), since the expected cost is the highest for all production points.

However, such approach has an essential drawback. Choosing one possible cost function value results in rejecting all other values and the plurality of input data is not considered.

Let us assume that a person making a decision has some information regarding a possible state of the environment and the corresponding cost functions.

State of the environment	Cost function			
θ_1	$c_1^1\left(x,\tau_i\right)$	$c_2^1\left(x,\tau_i\right)$	$\ldots$	$c_N^1\left(x,\tau_i\right)$
θ_2	$c_1^2\left(x,\tau_i\right)$	$c_2^2\left(x,\tau_i\right)$	$\ldots$	$c_N^2\left(x,\tau_i\right)$
$\ldots$	$\ldots$	$\ldots$	$\ldots$	$\ldots$
θ_M	$c_1^M\left(x,\tau_i\right)$	$c_2^M\left(x,\tau_i\right)$	$\cdots$	$c_N^M\left(x,\tau_i\right)$

Then, choosing a compromise function on the basis of decision-making (Truhaev 1981), makes it possible to some extent to take into account all the sets of cost functions. For such case the principle of insufficient reason can be used. It implies that in the case of absent reasons to consider some state of the environment to be more probable than any others, their a priori probabilities should be considered equal and the arithmetic average of all possible functions can be chosen as a compromise function.

$$c_i\left(x,\tau_i\right) = \frac{1}{M}\sum_{j=1}^{M} c_i^j\left(x,\tau_i\right), \ i=\overline{1,N}, \ x \in \Omega. \qquad (14)$$

Choosing a compromise function in accordance with Hurwitz criterion is another option in such situation:

$$c_i\left(x,\tau_i\right) = \alpha_i \min_{\theta_j} c_i^j\left(x,\tau_i\right) + \left(1-\alpha_i\right)\max_{\theta_j} c_i^j\left(x,\tau_i\right),$$

$$0 \le \alpha_i \le 1, \ i=\overline{1,N}, \ x \in \Omega. \qquad (15)$$

The coefficients α_i, $i=\overline{1,N}$ make it possible to take into account optimism-pessimism level according to the opinion of the person making a decision.

If a decision maker has the information about the probabilities of decision-making situations, it is possible to use an approach based on the criteria of decision-making under risk. Namely,

– choosing mathematical expectation as a feasible cost function:

$$c_i\left(x,\tau_i\right) = \sum_{j=1}^{M} p_j c_i^j\left(x,\tau_i\right), \qquad i=\overline{1,N}, \ x \in \Omega. \qquad (16)$$

– choosing the value corresponding to the most feasible environment state:

$$c_i\left(x,\tau_i\right) = c_i^{j^*}\left(x,\tau_i\right), \text{ где } p_{j^*} = \max_{\theta_j} p_j, \ i=\overline{1,N},$$

$$x \in \Omega. \qquad (17)$$

Other criteria can also be applied (Truhaev 1981).

When the compromise cost functions are chosen, the input task of infinite-dimensional mathematical programming with a multi-valued functional is reduced to a standard continual set partitioning problem, which has well developed methods and solving algorithms.

Let us formulate the solving algorithm for problem (5) – (8).

1. One of the criteria (12) – (17) is chosen considering existing information with regard to the decision-making situation.

2. The input set of objective functions is transformed into a single compromise function by applying the chosen criterion. In this way the initial multi-valued problem is reduced to a usual problem of continual set partitioning.

3. The obtained problem is solved by using the method of set partitioning from (Kiseleva & Shor 2005).

2.3 *Model problem and the analysis of the obtained results*

Supposing there is a set $\Omega = \left\{(x,y) \in R^2 | 0 \le x \le 10, \ 0 \le y \le 10\right\}$ of consumers of some homogeneous product. They are placed in the area Ω continuously. The initial coordinates of producers are given as $\tau_i = \left(\tau_i^x, \tau_i^y\right) = (0,0)$, $i=1,2,3$. Since the demand $\rho(x)$ for product at each point (x,y) of the set Ω is known, let us consider $\rho(x,y)=1$, $\forall (x,y) \in \Omega$ for simplicity and a set of feasible values of delivery cost function $c_i\left(x,y,\tau_i\right) = \left\{c_i^1\left(x,y,\tau_i\right); c_i^2\left(x,y,\tau_i\right); \ldots c_i^{M_i}\left(x,y,\tau_i\right)\right\}$, $i=\overline{1,3}$ from a producer τ_i to a client (x,y) according to the state of the environment θ_j be given as:

$$c_i^j\left(x,\tau_i\right) = \gamma_i^j \sqrt{\alpha_i^j\left(x-\tau_i^x\right)^2 + \beta_i^j\left(y-\tau_i^y\right)^2}, \ j=1,5.$$

The parameter values α_i^j, β_i^j, $i = \overline{1,3}$, $j = 1,5$ are shown in Table 1.

Table 1

State of environment	Delivery cost function								
	Producer 1			Producer 2			Producer 3		
	α_1^j	β_1^j	γ_1^j	α_2^j	β_2^j	γ_2^j	α_3^j	β_3^j	γ_3^j
θ_1	1	1	1	1	1	1	1	1	1
θ_2	0.5	1	1	1	1	2	1	1	2
θ_3	1	1.5	2	0.5	0.5	0.5	1	1	1
θ_4	0.5	1	2	1.5	3	2	2	3	1
θ_5	1	1	1	2	2	1	1	1	4

The producer's profit is supposed to depend only on transportation costs. The capacity of the producer i is defined by a total demand of the service clients and must not be higher than a given value b_i, $i = \overline{1,3}$: $b_1 = 25$, $b_2 = 60$, $b_3 = 50$. The area Ω is required to be divided into service zones by each of the producers, i.e. subsets Ω_1, Ω_2, Ω_3 to minimize the total costs on product delivery.

The following results were obtained by using the Hurwitz criterion with a constant $\alpha = 0.7$.

The producers' coordinates are:
τ_1 : (7.9098, 5.1205),
τ_2 : (3.1354, 2.1786),
τ_3 : (2.9033, 7.3819).

The optimal producers' capacities are 19.94, 38.25, 41.75. The functional value is 177.893.

The optimal partitioning of set Ω is shown in Fig. 1.

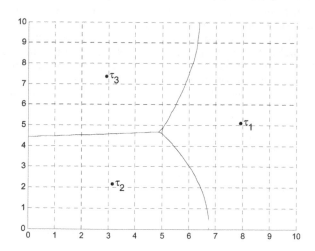

Figure 1. The optimal partitioning of set Ω by the Hurwitz criterion with a constant $\alpha = 0.7$.

By applying the Bayes criterion, the following result was obtained:

The producers' coordinates are:
τ_1 : (-31.1355, 15.1742),
τ_2 : (2.5073, 4.9982),
τ_3 : (7.5006, 4.9995).

The optimal producers' capacities are 0.0 50.0 50.0; The functional value is 269.020.

The optimal partitioning of set Ω is shown in Fig. 2.

In that case one of enterprises, namely τ_1, was not placed. Its capacity is 0.

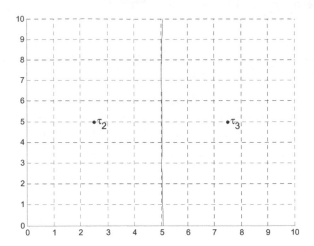

Figure 2. The optimal partitioning of set Ω by using Bayes criterion

4 CONCLUSIONS

The location problem under uncertainty is considered in this paper. Some of the appropriate mathematical models with multi-valued function are proposed. The algorithm solution based on a continual set partitioning method is proposed for this model. Experiments show that the proposed algorithm is feasible for providing the best results and can be used successfully. However, applying different criteria of decision making for transition from the multi-valued problem to a scalar one gives various compromise cost functions and, accordingly, different solutions of the original problem. Therefore, the next challenge is to examine the effectiveness of the solutions obtained.

Generalization of the considered problem will be infinite-dimensional location problems in fuzzy conditions. Mathematical models of such problems were formulated in the paper (Us 2010)

REFERENCE

Farahani R.Z., Hekmatfar M. 2009. *Facility Location: Concepts, Models, Algorithms and Case Studies.* Dordrecht: Heidelberg: London: New York: Springer.

Drezner Z., Hamacher. H. 2001. *Facility Location: Application and Theory.* Berlin: Springer.

Kiseleva E., Shor N. 2005. *Continuous optimal set partitioning: theory, algorithms, applications: Monography* (in Russian). Kiev: Naukova dumka.

Us S.A. 2013. *Application of the optimal set partitioning method to the problem of wireless network engineering.* CRC Press/Balkema: Taylor &Francis Group: Energy Efficiency Improvement of Geotechnical Systems – International Forum on Energy Efficiency: 175–181.

Corley H.W., Roberts S.D. 1972 *A Partitioning Problem with Applications in Regional.* Design Operations Research, Issue 20: 1010-1019.

Us S.A., Sadovnikova A.V. 1999. *Construction of fuzzy boundary in optimal set partitioning problem* (in Russian). Dnipropetrovs'k: DNU: Issues of applied mathematics and mathematical modeling: 151 – 159.

Us S.A. 2011. *Set partitioning problem with multi-valued objective functional* (in Russian). Dnipropetrovs'k: DNU: Issues of applied mathematics and mathematical modeling: 269 – 279.

Us S.A. 2011. *Application of statistical decision making to optimal set partitioning problem under uncertainty* (in Ukrainian). Dnipropetrovs'k: DNU: Collected research works of IX International Scientific Conference "Mathematical and Intelligent System Software" (MPZIS – 2011): 270 – 271.

Us S.A. 2012. *About one approach to infinite-dimensional problem solving of set partitioning with interval objective functional* (in Russian). Dnipropetrovs'k: DNU: Issues of applied mathematics and mathematical modeling: 254 – 260.

Kiseleva E.M. 2010. *Method of solving fuzzy set partitioning problem* (in Ukrainian). Dnipropetrovs'k: DNU: Issues of applied mathematics and mathematical modeling: 139 – 145.

Truhaev R.I. 1981. *Models of decision making under uncertainty* (in Russian). Moscow: Nauka.

Us S.A. 2010. *About optimal set partitioning problem under uncertainty* (in Russian). Dnipropetrovs'k: DNU: Issues of applied mathematics and mathematical modeling: 320–326.

© 2015 Taylor & Francis Group, London, ISBN 978-1-138-02804-3

Identification of stabilizing modes for the basic parameters of drilling tools

L. Meshcheriakov, L. Tokar & K. Ziborov
State Higher Educational Institution "National Mining University", Dnipropetrovs'k, Ukraine

ABSTRACT: The paper presents the results of identifying characteristics and properties of heteroscedastic nonlinear regression modes of conditional assessment of the technological parameter of mechanical diamond drilling speed dependence on the flushing liquid consumption rotation speed, and on rock cutting instrument loading

1 INTRODUCTION

All mining electromechanical industrial facilities, especially drilling complexes, are multiply connected, multivariable, dynamic, stochastic, nonlinear, and nonstationary systems with a great number of properties, which sophisticates greatly the process of their identification and control. However, it should be noted that generally property vector is not limited for any object of control, and the idea of "basic properties" is relative. Hence, synthesis of automated systems of identification and control of abovementioned objects stipulates the necessity to process huge data bases of vectors of input and output variables characterizing their technological and technical state. Maximum available state-of-the art information is required to develop a feasible identified model adequate to drilling facilities under consideration. On the basis of this, it is possible to develop optimum specific automated system of identification and control with the help of prediction model.

2 MAIN PART

The so-called epistemic models in electromechanical, mechanical, physical, chemical and other forms are developed in theoretical and experimental research and applied for sophisticated and structural drilling facilities both long-operating and newly-designed. Thus, it is possible to obtain diversified information body allowing selecting preliminary structure, information variables and approaches to identification of information features and parameters, which may help to single out other critical information factors. However, due to constructive complexity of drilling facilities, transition from epistemic models to information ones is insufficiently formalized which makes direct transition impossible.

Experimental research of technological and technical drilling parameters as shown in literature (Meshcheriakov 2009; 2007; 2007), resulted in nonlinear dependences of resource X of diamond crowns on drilling modes according to flushing load H, liquid consumption Q, and rotation speed of rock cutting tool ω. Information capacity of these dependences may be increased considerably by using conventionally probable moment any assessments proposed by (Meshcheriakov 2009). Such approach will improve the efficiency of control processes of drilling facilities due to increased data support, and will allow determining actual effects of such factors as drill column vibration frequency, alternation of different rocks, and ways of drilling parameters control. Their impact on rock cutting tools results in glazing, burning, and sticking, which worsens economic performance of drilling facilities.

Like any mathematical tool, methods of moment identification (Meshcheriakov 2009) are comprehensive and correct if only certain conditions are met. The best conditions providing theoretical feasibility of the methods are: independence of random variables within specific measurements from previous and next variables; standard distribution of input variable or state variable; and conditional dispersion stability of dependent output variable. Qualitative solution of control problem depends on the latter. In this context it is necessary to determine the degree of control object heteroscedasticity. Heteroscedastic effect may result on several factors: wear of equipment, errors and restrictions in signal measuring, changes in rocks composition and a number of other perturbing effects. Control objects manifest themselves as heteroscedastic in the process of nonlinear object linearization if a nonlinear object is represented as linear stochastic one where the first two moment conditional and unconditional characteristics coincide with corresponding moment characteristics of nonlinear objects. Research into such objects becomes more complicated than more accurate linear systems research. However, in many cases it results in significantly increase of input linear object or nonlinear object.

Analysis of known regressive models of technological and technical parameters of drilling process in terms of heteroscedasticity brings to light rather important information characteristics. Thus, nonlinear heteroscedastic regressive model of conditional mathematical expectation $M(V|Q)$ (Fig. 1, a) of dependence of technological parameter of mechanical drilling speed by means of diamond crowns V on flushing liquid consumption Q has a characteristic of considerable decrease along the conditional dispersion variable $\pm D(V|Q)$ within the zone of rational operating modes (230 – 250 l per minute), and increase at the boundaries of the working zone (100 and 300 l per minute). In accordance with conditional asymmetries $\pm A(V|Q)$ (Fig. 1, b), conditional kurtoses $\pm E(V|Q)$ (Fig. 1, c), and conditional variations $\pm Vr(V|Q)$ (Fig. 1, d), dependences of the parameters show increasing linear heteroscedasticity of characteristics within the working zone on $M(V|Q)$ (250-300 l per minute) minimum.

It is typical to observe minimization of dissipation of conditional asymmetries $\pm A(V|Q)$ and conditional kurtoses $\pm E(V|Q)$ regarding conditional mathematical expectations $M(V|Q)$ within the zone of rational operating modes 220-240 l per minute (Fig. 1, b and c), and increase in dissipation with decrease in flushing liquid consumption Q to 100 l per minute value. Alternatively, dependence of conditional variation $\pm Vr(V|Q)$ within the zone of rational operating modes increases (Fig. 1, d).

Nonlinear heteroscedastic regressive model of conditional mathematical expectation $M(V|\omega)$ (Fig.

2, a) presenting dependence of technological parameter of mechanical drilling speed diamond crowns V on rotational speed of rock cutting tool has matching $M(V|Q)$ characteristics on variable conditional dispersions $\pm D(V|\omega)$ (Fig. 2, a), conditional asymmetries (Fig. 2, b), conditional kurtoses $\pm E(V|\omega)$ (Fig. 2, c), and conditional variations $\pm Vr(V|\omega)$ (Fig. 2, d). Typical maximization of conditional kurtosis $\pm E(V|\omega)$ is observed with rational operating zones 220-240 l per minute (Fig. 2, c) as well as increase in values of conditional variations $\pm Vr(V|\omega)$ on the background of decreasing rotation speed of rock cutting instrument to 100 m per second (Fig. 2, d).

Fig. 3 shows basic characteristics of nonlinear heteroscedastic regressive models of conditional mathematical expectations $M(V|H)$ presenting dependence of technological parameter of mechanical drilling speed by means of diamond crowns V on loading parameters H of rock cutting instrument with conditional dispersions $\pm D(V|H)$, conditional asymmetries $\pm A(V|H)$, conditional kurtoses $\pm E(V|H)$, and conditional variations $\pm Vr(V|H)$. The general tendency decreasing dissipation of conditional dispersion $\pm D(V|H)$, conditional asymmetry $\pm A(V|H)$, conditional kurtosis $\pm E(V|H)$, and conditional variation $\pm Vr(V|H)$ with increased load on rock cutting instrument is. These values are stable within the zone of rational load upon rock cutting instrument when $M(V|H)$ is maximum (1050-1150 kgf). In the zone below the maximum value of mechanical drilling speed V, conditional dispersion $\pm D(V|H)$, conditional asymmetry $\pm A(V|H)$, and conditional variation $\pm Vr(V|H)$ have homoscedastic characteristics. In contrast, conditional kurtosis $\pm E(V|H)$ has alternating value (Fig. 3, c) within the zone.

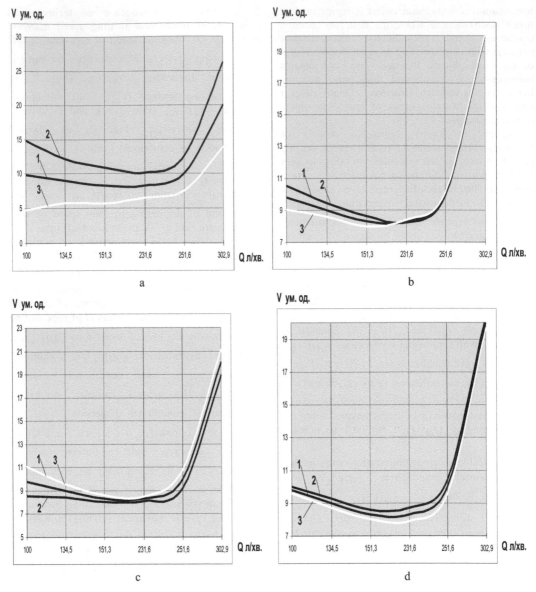

Figure 1. Nonlinear heteroscedastic regressive models of conditional mathematical expectations $M(V|Q)$ (1) with variable conditional dispersions $\pm D(V|Q)$ (a), conditional asymmetries $\pm A(V|Q)$ (b), conditional kurtoses $\pm E(V|Q)$ (c), and conditional variations $\pm Vr(V|Q)$ (d) related to the dependence of technological parameter of mechanical drilling speed by diamond crowns V on flushing liquid consumption Q.

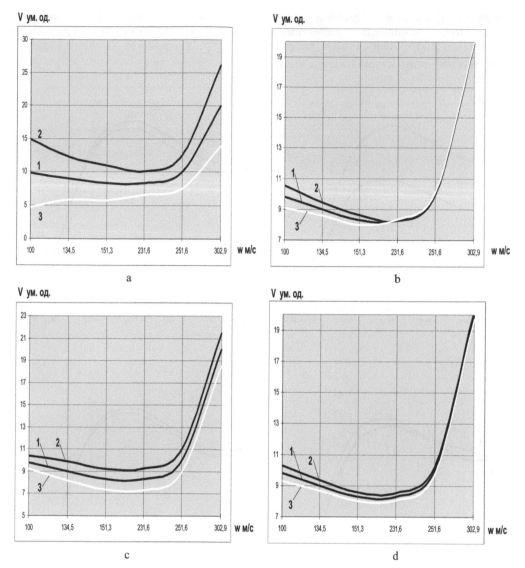

Figure 2. Nonlinear heteroscedastic regressive models of conditional mathematical expectations $M(V\,|\,\omega)$ (1) with variable conditional dispersions (a), conditional asymmetries $\pm A(V\,|\,\omega)$ (b), conditional kurtoses $\pm E(V\,|\,\omega)$ (c), and conditional variations $\pm Vr(V\,|\,\omega)$ (d) related to the dependence of mechanical drilling speed by means of diamond crowns V on the circular rotation velocity.

Fig. 4 shows nonlinear heteroscedastic regressive models of conditional mathematical expectations $M(X\,|\,Q)$ of drilling variable parameter per diamond crown X dependence on flushing liquid consumption Q (1) with conditional dispersions $\pm D(X\,|\,Q)$, conditional asymmetries , conditional kurtoses $\pm E(X\,|\,Q)$, and conditional variations $\pm Vr(X\,|\,Q)$. Maximum value of drilling variable parameter per diamond crown X is accompanied by maximum conditional dispersions $\pm D(X\,|\,Q)$ (Fig. 4, a). Its decrease is observed at the ends of operating modes zones regarding flushing liquid consumption Q. This characteristic is significant as well. The characteristic of conditional asymmetry $\pm A(X\,|\,Q)$ (Figure 4, b) having heteroscedastic

138

form with its minimization of deviation values departure from conditional mathematical expectations $M(X \mid Q)$ within a zone rational operating modes by means of typical decrease at the ends of modes zones flushing liquid consumption Q which provides interesting information for analysis.

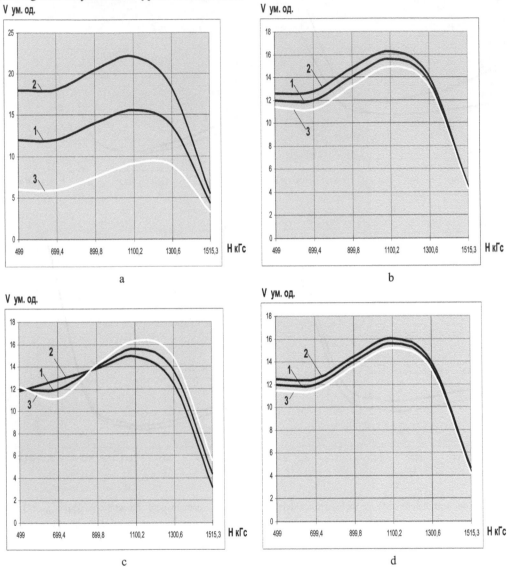

Figure 3. Nonlinear heteroscedastic regressive models of conditional mathematical expectations $M(V \mid H)$[1] with conditional dispersions $\pm D(V \mid H)$, conditional asymmetries $\pm A(V \mid H)$ (b), conditional kurtoses $\pm E(V \mid H)$ (c), and conditional variations $\pm Vr(V \mid H)$ (d) related to the technological parameter of mechanical drilling speed dependence on the rotation speed of the rock cutting tools

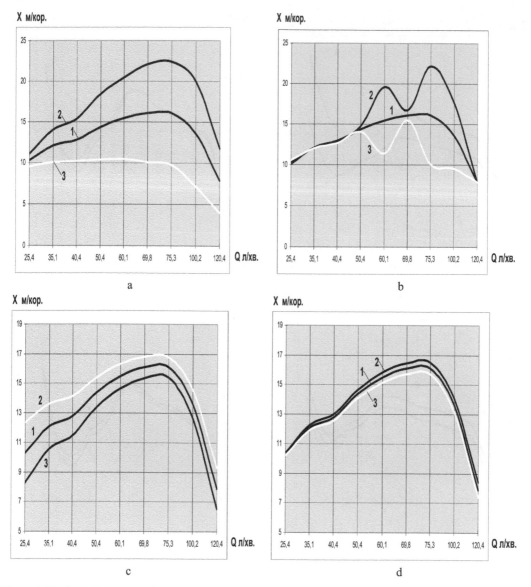

Figure 4. Nonlinear heteroscedastic regressive models of conditional mathematical expectations $M(X\,|\,Q)$ (1) with conditional dispersions $\pm D(X\,|\,Q)$ (a), conditional asymmetries $\pm A(X\,|\,Q)$ (b), conditional kurtoses $\pm E(X\,|\,Q)$ (c), and conditional variations $\pm Vr(X\,|\,Q)$ (d) of technical drilling parameter dependence per diamond crown X on flushing liquid consumption Q.

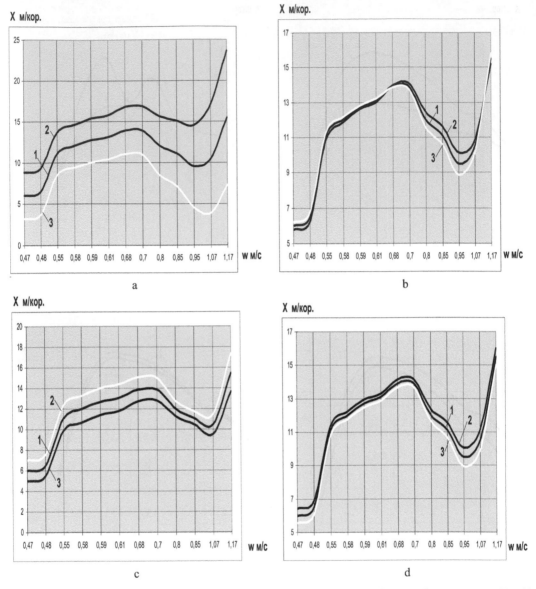

Figure 5. Nonlinear heteroscedastic regressive models of conditional mathematical expectations $M(X \mid \omega)$ (1) with conditional dispersions $\pm D(X \mid \omega)$ (a), conditional asymmetries $\pm A(X \mid \omega)$ (b), conditional kurtoses $\pm E(X \mid \omega)$ (c), and conditional variations $\pm Vr(X \mid \omega)$ (d) of technical drilling parameter per diamond crown X dependence on rotation speed ω of rock cutting instrument.

Dependence of conditional kurtosis $\pm E(X \mid Q)$ results from smaller divergence within the zone of growing conditional mathematical expectations $M(X \mid Q)$ up to the extreme point as well as from transition to homoscedastic mode with $M(X \mid Q)$ decrease within operating zone.

Dependence of conditional variation $\pm Vr(X \mid Q)$ is shown as typical decrease in departures from conditional mathematical expectations $M(X \mid Q)$ at the ends of operating zones, and departure maximization within rational operating modes.

Fig. 5 shows nonlinear heteroscedastic regressive models of conditional mathematical expectations

$M(X \mid \omega)$ of drilling parameter per diamond crown X dependence on rotational speed ω of rock cutting instrument with conditional dispersions $\pm D(X \mid \omega)$, conditional asymmetries $\pm A(X \mid \omega)$, conditional kurtosis $\pm E(X \mid \omega)$, and conditional variations $\pm Vr(X \mid \omega)$. True heteroscedastic characteristic (Fig. 5, a) stipulates twofold increase in conditional dispersion $\pm D(X \mid \omega)$ departure depending upon increase in rotational speed ω of rock cutting instrument within operating modes. Divergence of values of conditional asymmetries , conditional kurtosis $\pm E(X \mid \omega)$, and conditional variations $\pm Vr(X \mid \omega)$ from conditional mathematical expectations $M(X \mid \omega)$ is not significant; it increases only within the zone of rational operating modes. However, dependence data (Fig. 5) have five typical zones; each of them may be considered homoscedastic with adequate tolerance (0.47-0.48, 0.48-0.55, 0.55-0.1, 0.1-1.0, 1.0-1.11 m per second). In general, within each of the zones, different divergences of conditional dispersions $\pm D(X \mid \omega)$, conditional asymmetries $\pm A(X \mid \omega)$, conditional kurtosis $\pm E(X \mid \omega)$, and conditional variations $\pm Vr(X \mid \omega)$ characterize general dependence as heteroscedastic one with drastic change in direction.

3 CONCLUSIONS

Hence, identification and analysis of the conditional and instant factors of technological and technical parameters of drilling by means of diamond crowns allow discovering a number of significant information relations which reveal internal linear and nonlinear relationships of basic parameters of different operating modes of drilling facilities. Such approach broadens and deepens information support of automated systems of drilling facilities control resulting in improving accuracy and assurance of identifying operative technological and technical states of the latter. Besides, it makes automation of hole boring control more efficient.

REFERENCES

Meshcheriakov L.I. 2009. *Methods and models of identifying and controlling mining process systems: Monograph* (in Russian). Dnipropetrovs'k: National Mining University: 263.

Meshcheriakov L.I. 2007. *Intelligent technology of identifying resource of drilling facilities rock cutting instrument* (in Russian). Mining electromechanics and automation, Issue 79: 84–91.

Meshcheriakov L.I. 2007. *Intelligent diagnosis of drilling facilities* (in Russian). Dnipropetrovs'k: Collection of scientific papers of NMUU, Issue 29: 132–141.

Power Engineering, Control and Information Technologies in Geotechnical Systems – Pivnyak, Beshta & Alekseyev (eds)
© 2015 Taylor & Francis Group, London, ISBN 978-1-138-02804-3

The hydraulic impact and alleviation phenomena numeric modeling in the industrial pumped pipelines

A. Ghavrish & O. Shevtsova
State Higher Educational Institution "National Mining University", Dnipropetrovs'k, Ukraine

ABSTRACT: The issues of hydropercussion phenomena mathematical modeling in the industrial piping systems, with the pumps and dampeners, to determine the impact absorbers effectiveness on the amplitude-frequency characteristics of these hydromechanical systems are considered.

1 INTRODUCTION

Method of calculating the transient and frequency characteristics of the pipeline that contains the pump and the dampener, is based on nonlinear mathematical model. Simulation of overlapping stream with using industrial valves is provided by introducing the exponential law of diminishing cross-sectional area of the pipeline. The basis of calculation is the method of characteristics applied to the simplified Navier-Stokes equations. The resulting nonlinear differential equations are solving by using the finite difference method of first order.

2 ANALYSIS OF EXISTING ACHIEVEMENTS AND PUBLICATIONS

In this article the problem of the damping devices impact estimation to smooth the peak amplitudes of pressure in the hydraulic fluids industrial pipelines under the simulated conditions of hydropercussion processes when the modified method of characteristics are used are considered (Dorch, Wood & Lightner 1966). The software part of the project was implemented using the built-in programming language of Matlab software package (The MathWorks 2002).

3 FORMULATION OF THE RESEARCH GOALS AND OBJECTIVES

Purpose of work is to develop the mathematical model of hydrodynamic flow of the working fluid in a long discharge pipe considering slackness and roughness of its walls in imitation of a sharp stop this flow by closing the valve at its distal end with the further implementation of this model in a software package based on Matlab package (actually it is the task of Computational Fluid Dynamics - CFD). On the basis of the mathematical model the task to create a software package, which includes user-friendly graphical interface allows you to use the results of this development in the form of an executable .exe program. In the future, this program is intended for estimations of the damping devices effectiveness of different parametric configurations on the long lasting industrial pressurised pipeline systems, the characteristics of which input at working procedure through the Windows OS based PC as a workstation by trained personnel of the damping devices manufacturer.

These developments have been carried out under the framework of a scientific mission to the Liquid Dynamics int'l inc company, headquartered in Wilmington area, NC – World wide manufacturer and exporter of damping devices, and on the instructions and order of Software Engineering Cathedra of Dnepropetrovsk National Mining University, Ukraine.

4 THE MAIN PART OF THE RESEARCH REPRESENTATION

4.1 Method of characteristics.

To solve this problem in this paper the method of characteristics (Bakar & Firoz 2002.) has been used. The method of characteristics converts partial differential equations, for which the solution can't be written in general terms (as, for example, the equations describing the fluid flow in a pipe) into the equations in total derivatives. The resulting nonlinear equations can then be integrated using the methods of using the equations of finite differences.

4.2 Equations of motion

Hydraulics equations that embody the principles of conservation of angular momentum and continuity in the one-line pipe, respectively, are as follows:

$$+\begin{cases} \dfrac{P_x}{\rho}+VV_x+V-g_t\cdot\sin a+\dfrac{fV|V|}{2D}=0 \\ PP_t+P_xV+\rho\cdot a^2\cdot V_x=0 \end{cases}\Big|\times\lambda \qquad (1)\,(2)$$

These equations can be combined with the unknown factor of λ and obtain the equation:

$$\lambda\left[P_x\left(V+\dfrac{1}{\lambda\rho}\right)+P_t\right]+\left[V_x\left(V+\rho\cdot a^2\lambda\right)+V_t\right]-g\cdot\sin a+\dfrac{fV|V|}{2D}=0 \qquad (3)$$

The arbitrary choice of two different values of λ give two independent equations in the variables $P(x,t)$, $V(x,t)$, is equivalent to (1) and (2). With a suitable choice of λ the simplification is possible. In particular, since P and V are functions of x and t, then if we assume that x - a function t, then:

$$\dfrac{dP(x,t)}{dt}=\dfrac{\partial P}{\partial x}\dfrac{dx}{dt}+\dfrac{\partial P}{\partial t}\dfrac{dt}{dt}=P_x\dfrac{dx}{dt}+Pt$$

$$\dfrac{dV(x,t)}{dt}=\dfrac{\partial V}{\partial x}\dfrac{dx}{dt}+\dfrac{\partial V}{\partial t}\dfrac{dt}{dt}=V_x\dfrac{dx}{dt}+V_t \qquad (4)$$

If $\quad \dfrac{dx}{dt}=V+\dfrac{1}{\lambda\rho}=V+\rho a^2\lambda \qquad (5),$

then equation (3) becomes an ordinary differential equation:

$$\lambda\dfrac{dP}{dt}+\dfrac{dV}{dt}-g\sin a+\dfrac{fV|V|}{2D}=0 \qquad (6)$$

Solving (5), we obtain:

$$\lambda=\dfrac{1}{\rho a},\quad \dfrac{dx}{dt}=V+a \text{ - downstream,}$$

$$\lambda=\dfrac{1}{\rho a},\quad \dfrac{dx}{dt}=V+a \text{ - upstream} \qquad (7)$$

Substituting equation (7) (6), we obtain a system of total differential equations:

$$\dfrac{dP}{dt}+\rho a\dfrac{dV}{dt}-\rho ag\sin a+\rho a\dfrac{fV|V|}{2D}=0, \qquad (8)$$

$$\dfrac{dx}{dt}=V+a, \qquad (9)$$

$$\dfrac{dP}{dt}-\rho a\dfrac{dV}{dt}+\rho ag\sin a-\rho a\dfrac{fV|V|}{2D}=0, \qquad (10)$$

$$\dfrac{dx}{dt}=V-a \qquad (11)$$

4.3 Finite-difference scheme

For solving of the nonlinear equations (8) - (11) a finite difference method is used.

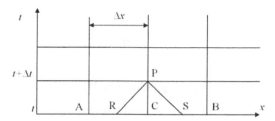

Figure 1. Spatial - temporal grid

Spatial - temporal grid (Fig. 1) describes the state of the liquid at various points in the pipeline at time t and t + Δt. Pressure and velocity at points A, C and D, which correspond to time t, are known either from the previous step, or from data on the steady flow. States at R and S correspond to the time t and should be calculated from the values at points A, C and B. State at the point P corresponds to the time t + Δt is determined from equations (8) - (11). Let us to write the equations (8) - (11) in finite differences and multiply it by an increment of time:

$$P_p-P_R+\rho a_R(V_p-V_R)-\rho a_Rg\sin a(t_p-t_R)+\rho a_R\dfrac{fV_R|V_R|}{2D}(t_p-t_R)=0, \qquad (12)$$

$$(x_p-x_R)=(V_R+a_R)(t_p-t_R), \qquad (13)$$

$$P_P-P_S+\rho a_S(V_S-V_S)-\rho a_Sg\sin a(t_P-t_S)+\rho a_S\dfrac{fV_S|V_S|}{2D}(t_P-t_S)=0, \qquad (14)$$

$$(x_P-x_S)=(V_S+a_S)(t_P-t_S), \qquad (15)$$

We have used a constant time step - a special time interval.

4.4 Special time interval.

We will write the equations (12) and (14) as:

$$P_p=C_p-\rho a_RV_P \qquad (16)$$

144

$$P_P = C_M + \rho a_S V_P \qquad (17),$$

where

$$C_P = P_R + \rho a_R V_R \left(1 + \frac{g}{V_R} \Delta t \sin a - \frac{f \Delta t |V_R|}{2D}\right) \qquad (18)$$

$$C_M = P_S - \rho a_S V_S \left(1 + \frac{g}{V_S} \Delta t \sin a - \frac{f \Delta t |V_S|}{2D}\right) \qquad (19)$$

From Fig. 1 and (13):

$$\frac{x_C - x_R}{x_C - x_A} = \frac{V_C - V_R}{V_C - V_A},$$

$$(20) \Rightarrow V_R = V_C - \frac{x_C - x_R}{x_C - x_A}(V_C - V_A)$$

$$x_C - x_R = x_P - x_R = (V_R + a_R)(t_P - t_R)$$

We substitute this expression in (20):

$$\frac{(V_R + a_R)(t_P - t_R)}{x_C - x_A} = \frac{V_C - V_R}{V_C - V_A}, \Rightarrow$$

$$\Rightarrow [V_R(t_P - t_R) + a_R(t_P - t_R)](V_C - V_A) =$$
$$= (V_C - V_R)(x_C - x_A),$$

$$V_R(t_P - t_R)(V_C - V_A) + a_R(t_P - t_R)(V_C - V_A) =$$
$$= V_C(x_C - x_A) - V_R(x_C - x_A),$$

$$V_R[(t_P - t_R)(V_C - V_A) + x_C - x_A] = V_C(x_C - x_A) - a_R(t_P - t_R)(V_C - V_A),$$

$$V_R = \frac{V_C(x_C - x_A) - a_R(t_P - t_R)(V_C - V_A)}{(t_P - t_R)(V_C - V_A) + x_C - x_A} \qquad (*)$$

Divide (*) by $(x_C - x_A)$, given that
$t_P - t_R = \Delta t$, $x_C - x_A = \Delta x$:

$$V_R = \frac{V_C - a_R \frac{\Delta t}{\Delta x}(V_C - V_A)}{1 + \frac{\Delta t}{\Delta x}(V_C - V_A)} = \frac{V_C - \xi_R(V_C - V_A)}{1 + \theta(V_C - V_A)} \qquad (21)$$

$$\theta = \frac{\Delta t}{\Delta x} \qquad (22),$$

$$\xi_R = a_R \frac{\Delta t}{\Delta x} = \theta a_R \qquad (23)$$

Similarly, from Fig. 1 and (15) shows:

$$\frac{x_C - x_S}{x_C - x_B} = \frac{V_C - V_S}{V_C - V_B},$$

$$x_C - x_S = x_P - x_S = (V_S + a_S)(t_P - t_S)$$

Combine these two expressions: :

$$\frac{(V_S + a_S)(t_P - t_S)}{x_C - x_B} = \frac{V_C - V_S}{V_C - V_B}, \Rightarrow$$

$$\Rightarrow [V_S(t_P - t_S) - a_S(t_P - t_S)](V_C - V_B) =$$
$$= (V_C - V_S)(x_C - x_B),$$

$$V_S(t_P - t_S)(V_C - V_B) - a_S(t_P - t_S)(V_C - V_B) =$$
$$= V_C(x_C - x_B) - V_S(x_C - x_B),$$

$$V_S = \frac{V_C(x_C - x_B) + a_S(t_P - t_S)(V_C - V_A)}{(t_P - t_S)(V_C - V_B) + (x_C - x_B)} \qquad (**)$$

Given that $t_P - t_S = \Delta t$, $x_C - x_B = -\Delta x$ divide
(**) in $(x_C - x_B)$

$$V_R = \frac{V_C \frac{x_C - x_B}{x_C - x_B} + a_S \frac{t_P - t_S}{x_C - x_B}(V_C - V_B)}{\frac{x_C - x_B}{x_C - x_B} + \frac{t_P - t_S}{x_C - x_B}(V_C - V_B)} = \qquad (24)$$

$$= \frac{V_C - \xi_S(V_C - V_B)}{1 + \theta(V_C - V_B)}$$

Similarly,

$$\frac{x_C - x_R}{x_C - x_A} = \frac{P_C - P_R}{P_C - P_A} \Rightarrow P_R = P_C - \frac{x_C - x_R}{x_C - x_A}(P_C - P_A) \qquad (***)$$

$$\frac{x_C - x_R}{x_C - x_A} = \frac{x_P - x_R}{x_C - x_A} \quad \text{from Fig. 1. Substitute in}$$
this expression (13)

$$\frac{x_C - x_R}{x_C - x_A} = \frac{(V_R + a_R)(t_P - t_R)}{x_C - x_A} = (V_R + a_R)\frac{\Delta t}{\Delta x} + a_R \frac{\Delta t}{\Delta x} =$$
$$= V_R \theta_R + \xi_R$$

Substituting this expression in (***), obtain:

$$P_R = P_C - (V_R \theta_R + \xi_R)(P_C - P_A) \qquad (25)$$

Similarly,

$$\frac{x_C - x_S}{x_C - x_B} = \frac{P_C - P_S}{P_C - P_B} \Rightarrow P_S = P_C +$$
$$+ \frac{x_C - x_S}{x_B - x_C}(P_C - P_B) \qquad (****),$$

145

$$\frac{x_C - x_S}{x_B - x_C} = -\frac{x_C - x_S}{x_S - x_B} = -\frac{x_P - x_S}{x_C - x_B}.$$

We substitute this expression in (15):

$$\frac{x_C - x_S}{x_B - x_C} = -\frac{(V_S + a_S)(t_P - t_S)}{x_C - x_B} =$$

$$= (V_S - a_S)\frac{(t_P - t_S)}{x_B - x_C} = V_S\frac{\Delta t}{\Delta x} - a_S\frac{\Delta t}{\Delta x} = V_S\theta_S - \xi_S$$

We substitute this expression into

(****): $P_S = P_C + (V_S\theta_S - \xi_S)(P_C - P_B)$ (26)

To save the convergence of these equations imply satisfaction with the Courant conditions:

$$\xi \le \frac{a}{V + a}$$ (27).

These conditions imply that in Fig. 1 points R and S are located between points A and B. Solving the equation (16) and (17) with the aR=aS we get the pressure at point P:

$$P_P = \frac{C_P + C_M}{2}$$ (28)

To calculate the rate of VP can be any of the equations (16) and (17). This completely determines the state at all interior points of the pipeline. Note the use of linear interpolation of the pressure and velocity of the liquid in the pipeline. To maintain accuracy in the calculation of nonlinear systems, values Θ and ξ must satisfy the Courant inequalities that involve interpolation only a small step of the grid.

Thus, the problem of flow in the pipeline is completely solved for interior points, but there remains the problem of establishing the boundary conditions at the endpoints, in which neither the CP nor the CM is unknown.

At each end of the pipeline is only one of a pair of equations, i.e. equation (16) or (17). At the entrance to the pipeline is using the equation (17) and output from the pipeline - equation (16) (see Fig. 2).

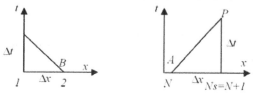

Figure 2. Spatially-time grid for the boundary conditions at the constant cross section pipeline ends.

In order to determine the pressure and velocity at the ends of the pipeline, it is necessary to bring the auxiliary equations (boundary conditions), defined by the conditions at the ends of the pipeline.

If the pressure at the inlet or outlet of the pipeline is a known function of time F (t), then this relation can be combined with equation (16) or (17) to determine the status of a boundary point.

A known pressure at the inlet to the pipeline:

$$P_U = F(t), P_D = P_U, P_D = C_M + \rho a_D V_D => V_D =$$
$$= \frac{F(t) - C_M}{\rho a_D}$$ (29)

A known pressure at the outlet of the pipeline:

$$P_D = F(t), P_U = P_D, P_U = C_P + \rho a_U V_U => V_U =$$
$$= \frac{C_P - F(t)}{\rho a_U}$$ (30)

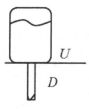

Fig. 3. Diagram illustrating the change in the cross section of the pipe.

The expression of continuity for incompressible fluid

$$A_U V_U = A_D V_D$$ (31),

where AU, AD - cross-sectional area of pipelines.

Assuming the absence of energy dissipation (i.e., the absence of losses) at the interface of pipelines and equating the full of pressure on each side of the junction, we obtain:

$$P_U = C_P - \rho a_U V_U,$$

$$P_D = C_M - \rho a_D V_D,$$

$$P_U + \frac{1}{2}\rho V_U^2 = P_D + \frac{1}{2}\rho V_D^2,$$

$$\frac{1}{2}\rho V_D^2 + C_M + \rho a_D V_D - \frac{1}{2}\rho V_U^2 - C_P + \rho a_U V_U = 0, \text{(32)}$$

$$\frac{1}{2}\rho V_D^2 + C_M + \rho a_D V_D - \frac{1}{2}\rho\left(\frac{A_D}{A_U}\right)^2 - V_D^2 + C_P +$$

$$+ \rho a_U \frac{A_D}{A_U} V_D = 0,$$

146

$$V_D^2(1-\beta^2)+2(a_D+\beta a_U)V_D+\frac{2}{\rho}(C_M-C_P)=0$$

Or $A\cdot V_D^2+B\cdot V_D+C=0a$ (33),

where

$$\beta=\frac{A_D}{A_U}$$

To obtain a positive rate of flow is necessary to use the positive square root of the formula:

$$V_D=\frac{-B+\sqrt{B^2-4AC}}{2A}$$ (34)

Substituting (34) into the (31), (16), (17) we obtain all the necessary quantities at the interface of the pipelines.

The only assumption made was the neglect of energy dissipation at the junction. This assumption is quite reasonable for the case of relatively low velocity of fluid flow in the piping systems under consideration, as well as for the piping systems with a streamlined shape.

The equation of flow through the diaphragm has the form:

$$V_D=\tau\sqrt{P_D-P_U}\,,$$ (35)

$$P_U=C_P-\rho a_U V_U, P_D=C_M+\rho a_D V_D, A_U V_U=A_D V_D$$

Where VD determined from the quadratic equation with coefficients

$A=1,$

$$B=\tau^2\rho(\beta a_U+a_D), c=\tau^2\left(C_M-C_P\right)$$

Other unknowns are easily found from (31), (16), (17).

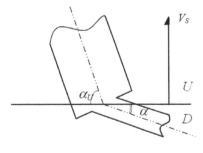

Figure 4. The joints scheme of various sections of pipelines

Fig. 4 shows the relative motion between the two elements of the system. If the size and orientation of the two elements are not identical, in general, in the place of joining of these elements will be some accumulation or exudation of liquid associated with a relative displacement of elements.

Equating the rate of accumulation of fluid flowing, and the difference between the expenditure of the effluent, we obtain:

$$A_U V_U - A_D V_D =$$
$$= -V_S\left(A_U \sin a_u - A_D \sin a_D\right)$$ (36)

If we assume that the velocity of the design section VS directed vertically, while the fluid velocity VU and VD directed along the axis of the pipe, oriented at an angle αU and αD, from the equations (16) and (17) $P_U=P_D$,

$$V_U=\beta V_D-V_S\left(\sin a_u-\beta\sin a_D\right)$$ (37)

The action of the pump can be approximated by the pressure jump in the value of ΔP, if the size of the pump is small in comparison with other elements of the system. The boundary conditions in this case would be the same as in the case when the pressure is known:

$$P_U=C_P-\rho a_U V_U, P_D=C_M+\rho a_D V_D, P_D=$$
$$= P_U+\Delta P, A_U V_U = A_D V_D,$$

$$C_M+\rho a_D V_D=C_P-\rho a_U\beta V_D+\Delta P =>$$

$$V_D=\frac{C_P-C_M+\Delta P}{\rho(a_D+\beta a_U)}$$ (38)

The damper is a concentrated yielding at a docking site of two elements.

Since the damper is a pressure accumulator (total energy and the amount of liquid in it are the function of pressure), the change in pressure in it is described by a differential equation. Yielding of the damper is given by:

$$b'=\frac{dVol}{dP}$$ (39)

As in the case of relative motion, the rate of accumulation of fluid in the damper is equal to the difference between the inlet and outlet pressure:

$$V_U A_U - V_D A_D = \frac{dVol}{dt} = \frac{dVol}{dP}\frac{dP}{dt} = b'\frac{dP_U}{dt},$$

$$\frac{dP_U}{dt} = \frac{V_U A_U - V_D A_D}{b'}$$ (40)

5 THE RESULTS OF THE CALCULATIONS

Implementation of this mathematical model to that shown in Fig. 5 a long pipeline diagrams was carried out by creation and debugging a software using a complex algorithm as an operation and interaction of software modules.

Since one of the problems was the task of creating the comprehensive software with its user-friendly graphical user interface (GUI), it was decided to carry out a set of numerical values of the original data through pop-up windows, which contains the information about the current input parameters with all the necessary explanations and tips. See Fig 6.

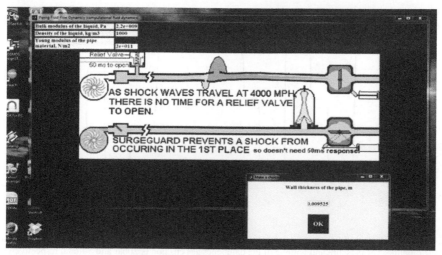

Figure 5. GUI look at the dataset.

Fig. 6 shows the GUI while setting the next value of the current parameter.

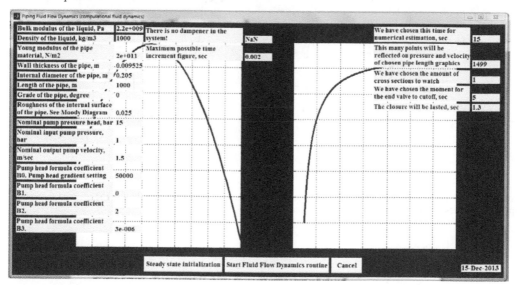

Figure 6. GUI look at the computation procedure. In the boxes the numerical values of the system parameters for which the current calculation is conducting are displaying.

148

Following are the results of the trial calculations for water pipe length of 3000 m, with a wall thickness of 9.525mm steel and an inner diameter of 205 mm. Rated pump head 15bar at nominal input pressure of 1 bar and a nominal flow rate at the outlet of the pump 1.5m/s (at a steady flow). The mass flow rate of the working fluid was 27.8 kg/sec (at a steady flow). At the end of the pipe before the valve was installed damper. The flap at the end of the pipeline began to close on 5 second. Response time to complete closure flap is 1.3 seconds.

Figs. 7, 8 and 9 show the graphs of changes in the instantaneous pressure and the particle velocity of the working fluid (water in this calculation) in the nodal point section of spatio-temporal grid nearest to the throttle duct for the three variants of the dampener parameters which is installed at the end of the pipeline.

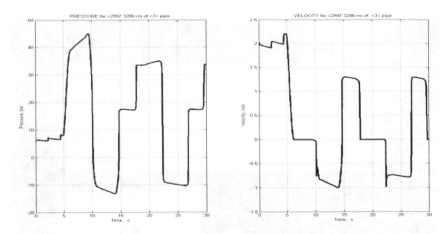

Figure 7. Figure of instantaneous changes in pressure and particle velocity of the working fluid in the selected section of the pipeline. The pressure and volume of the dampener gas cushion is 20bar and 5 liters, respectively

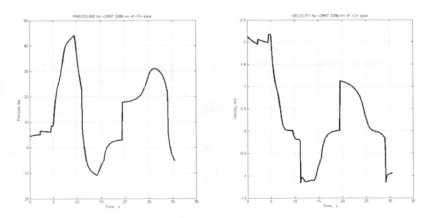

Figure 8. Figure instantaneous changes in pressure and particle velocity of the working fluid in the selected section of the pipeline. The pressure and volume of a gas cushion is 20 bar and 50 liters, respectively.

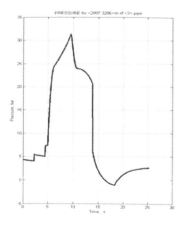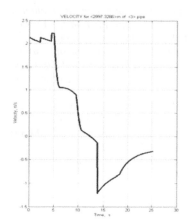

Figure 9. Figure instantaneous changes in pressure and particle velocity of the working fluid in the selected section of the pipeline.

Pressure and volume of a gas cushion is 20 bar and 250 liters, respectively.

As can be seen from Figs. 7, 8 and 9 graphs, obtained for the pump and piping configurations for different values of the volume of the gas cushion damping device, when sequentially building a gas cushion volume 5Lt, 50Lt, 250Lt, respectively the oscillation frequency of the working fluid with the elastic part of the steel pipe wall varies from 0.1 Hz through 0,067 Hz and at the value of the gas cushion 250Lt the flow after starting the unit spike of pressure, goes to be "aligned" in close to steady flow since 25 seconds. It can be concluded that at a pressure of a gas cushion equal to 20 bar, the volume of gas cavity damper should be used 250Lt to "smooth" flow in the pipeline and almost complete suppression of the hydraulic impact phenomena spread in a liquid medium for a given parameters combination such a long pipeline.

6 CONCLUSIONS

1. As a result of the completed research the familiarization with the methods of mathematical modeling of hydro-mechanical processes was implemented by using the method of characteristics applied to the modified nonlinear Navier-Stokes differential equations.

2. The proposed approach for the numerical integration of the Navier-Stokes equations on the spatio-temporal grid by the method of characteristics had allowed to input practical computing parameters of the working fluid in long pipelines, with taking into account the boundary conditions for the butt section of the pipeline, which determine the behavior of pumps, diaphragm valves, dampener devices connected to these sections.

3. The mathematical model for predicting of the peak pressure amplitudes caused by various inhibitory factors that lead to the return of the mass flow, and then with its use, program computer's complex were developed and debugged. For easier use of the program complex was created a graphical users interface. The program is finished to working condition and has been used as simulation software for the test workstation based on the PC with using the operating systems Windows Vista and Windows 7.

4. Implemented calculations have been showing the complete fitness of the results of these developments for the solution the main challenge goal of the project by correct chosen parameters of the alleviation dampener to get proper and smooth fluid flow in the pumped industrial pipelines and suppress the hydropercussion (water-hummer) phenomena. The water hummer effect has been numerically simulated by sudden overlap of flow section of the pipeline and the suppress of the arised pressure head was numerically simulated by using the dampener mounted to the distal end of the pipeline.

REFERENCES

Dorch R.G., Wood D.J., Lightner C. 1966. *Distributed parameters analysis of pressure and flow disturbances in rocket propellant feed systems.* Washington DC: NASA Technical Note: 54.

MATLAB Compiler, The Language of Technical Computing: - Sixth printing, Revised for Version 3.0. 2002. Natick: The MathWorks: 274.

Bakar R.A. & Firoz T. 2002. *Simulation of transient flows in a hydraulic system with a long liquid line.* American Journal of Applied Sciences, Issue 5, Vol. 2: 5.

Chaudhry M.H., Hussaini M.Y. 1985. *Second-order explicit finite-difference schemes for waterhammer analysis.* Journal of Fluids Engineering, Issue 107: 523-529.

Wylie E.B., Streeter V.L. 1978. *Fluid Transients.* New York: McGraw-Hill.

Power Engineering, Control and Information Technologies in Geotechnical Systems – Pivnyak, Beshta & Alekseyev (eds)

One way to solve problems of multi-zone dynamics models identification.

L. Koriashkina, A. Pravdivy
Oles Honchar Dnipropetrovs'k National University, Dnipropetrovs'k, Ukraine

A. Cherevatenko
State Higher Educational Institution "National Mining University", Dnipropetrovs'k, Ukraine

ABSTRACT: The paper presents a class of inverse problems for systems of differential equations in which the piecewise continuous right-hand side of the differential equation must be restored. Problems of dynamical systems identification are formulated in terms of the continuous optimal set partitioning (OSP) problems theory and are solved by the methods of nondifferential optimization, part of which is Shor`s r-algorithm.

1 INTRODUCTION

Many technical facilities such as thermal machines, electric motors, gyroscopic systems, moving objects, in conditions of the real operation, can significantly modify the components of the dynamics model. It may be caused by changes in the properties (thermal, electromagnetic) of materials as a result of changes in the parameters of an object's state, or in the conditions of its functioning. Dynamic regimes of such facilities or systems can be described precisely enough by the sequential use of several linear differential equations (Matveykin 2007, Filippov 1985). For this purpose, a set of all possible system states is partitioned into several stages or zones so that, within a single stage, dynamics would be described with sufficient accuracy by one (in the general case vector) linear differential equation.

In the case of separation of N zones, general dynamic model represents a set of N ordinary differential equations

$$\dot{x}(t) = A_1 \cdot x(t) + B_1 \cdot u(t), \quad y(t) \in \Omega_1,$$
$$\cdots \qquad\qquad\qquad\qquad\qquad (1)$$
$$\dot{x}(t) = A_N \cdot x(t) + B_N \cdot u(t), \quad y(t) \in \Omega_N,$$

where $x(t)$ – phase coordinate vector; $u(t)$ – control parameters vector; $A_j, B_j, j = \overline{1, N}$ – matrices of the system parameters; $y(t)$ – output (observable) variable, usually one of the phase coordinates, assuming values from some feasible set Ω; subsets $\Omega_1, \ldots, \Omega_N$ that constitute a partition of Ω, determine N possible zones of the system functioning. For example, if $y = x_1$ then $\Omega_1, \ldots, \Omega_N$ – partition of the set Ω of the first phase system coordinate possible values into N subsets.

In general case vector of the phase variables in (1) may have different dimensions at different stages. If the dimension of vectors x and u for all stages (zones) is the same, the system of equations can be represented by one differential equation with discontinuous right-hand side:

$$\dot{x}(t) = \begin{cases} A_1 x(t) + B_1 u(t), & y(t) \in \Omega_1, \\ \cdots \\ A_N x(t) + B_N u(t), & y(t) \in \Omega_N. \end{cases} \qquad (2)$$

This paper presents class of inverse problems for systems of type (2) in which the right-hand side of the differential equation (piecewise continuous function in its domain (Kiseleva 2013)) must be restored. This class is characterized by the fact that all problems can be formulated in terms of the theory of continuous dynamic OSP problems (Shor 1979) as problems of partitioning the continuum set in the best (in some sense) way and solved by corresponding methods. The mathematical theory of continuous problems of OSP is based on a common approach, which consists of reducing the original infinite optimization problems to non-smooth usually finite-dimensional optimization problems, which are numerically solved by modern effective non-differentiable optimization methods (Bessonov 1989).

2 PROBLEM STATEMENT

Let us consider the particular case of multi-zone dynamics model (2):

$$\begin{cases} \dot{x}_1 = x_2; \\ \dot{x}_2 = a_1^i x_1 + a_2^i x_2 + b^i f_i(v_i, t), \\ \qquad x_1 \in \Omega_i, i = \overline{1, N}. \end{cases} \qquad (3)$$

Here the state of an object is characterized by two phase coordinates, $\Omega_1, \ldots, \Omega_N$ – the sets of values of the system first phase coordinate. They define the possible states (zones) of the system operator; $f(v_i, t)$ – known function, continuous in each of its variables; $a_1^i, a_2^i, b^i, v_i, i = \overline{1, N}$ – given (or not) parameters.

The direct problem for the system (3) is formulated as follows: according to known boundaries between zones $\Omega_1, \Omega_2, \ldots, \Omega_N$ and given values of ' parameters $a_1^i, a_2^i, b^i, v_i, i = \overline{1, N}$ it is needed to find a solution of (3) subject to the condition:

$$x_1(0) = \hat{x}_1, \quad x_2(0) = \hat{x}_2. \qquad (4)$$

Under the system identification we will understand a refinement of the object model or the process (3) based on experimental data, i.e. determining the unknown boundaries between operation zones $\Omega_1, \ldots, \Omega_N$, as well as parameters $a_1^i, a_2^i, b^i, v_i, i = \overline{1, N}$ (if they are not set) under the terms (4) and aposteriori information about the values of first phase coordinate in some time points t_k:

$$x_1(t_k) = \tilde{x}_1(t_k), k = \overline{1, K}. \qquad (5)$$

The formulated problem of dynamic systems identification belongs to the class of inverse problems for ordinary differential equations. The theory of inverse problems for differential equations is intensively developing now which is caused to a considerable extent by the necessity to develop mathematical methods for solving a wide range of important applied problems associated with processing and interpretation of the observations

(Grop 1979; Isermann 2010). Despite the fact that the intensive study of inverse problems started relatively recently, a large number of significant results has already been received in this area. Nowadays, several schools are formed in regard to both variety of applications and types of mathematical inverse problems statements. The number of scientific publications on the theory of inverse problems and its applications is very large. Many of these results are reflected in the books, where general issues are considered as well as special sections devoted to specific research directions (Bessonov 1989; Grop 1979; Vatulyan 2007).

As is well known, one of the approaches to solving identification problems of dynamic systems is to reduce it to the problem of optimal control and apply to the solution of the latter, for instance, the Pontriagin maximum principle or dynamic programming techniques. But during the formulation of optimal control problems by complex processes and phenomena it can be sometimes difficult to control adequately and construct a satisfactory measure of quality achievement for a variety of objectives. Therefore, the development of specialized methods for solving inverse problems for differential equations with switched right-hand side, of their algorithms and software is currently the topical research direction.

In this paper we propose to apply methods of the theory of continuous OSP problems (Kiseleva 2013) for solving identification problems of complex multistage processes. Problems and methods of their solution presented in the paper synthesize fundamental principles of the theory of continuous OSP problems and the theory of inverse problems for lumped parameters systems.

3 FORMULATION OF THE IDENTIFICATION PROBLEM IN TERMS OF THE THEORY OF CONTINUOUS OSP PROBLEMS

The inverse problem for system (3) is formulated on the assumption that the boundaries of switching zones are not known. The values of x_1 are observed only at certain time moments of the interval $\overline{T} = \{t : 0 \le t \le T\}$. It is necessary to determine boundaries between the zones in system (3) and probably parameters of right sides functions $a_j^i, b^i, v_i, i = \overline{1, N}$ (if they are not set) from the known values of the first phase coordinate $\tilde{x}_1$ in the points of the interval $\overline{T}$ (using conditions (5)).

Let us formulate mathematically the problem of multi-zone dynamic system identification, involving the basic concepts of the theory of continuous OSP problems.

Suppose the measurements results of input and output variables of the object are known and the number N of the object operation zones $\Omega_1,\ldots,\Omega_N$ is set (the set Ω of values of the first system phase coordinate). Let us introduce the following notation:

- $P_N(\Omega)$ – class of all possible partitionings of the set Ω at N subsets $\Omega_1,\ldots,\Omega_N$:

$$P_N(\Omega) = \{\overline{\omega} = (\Omega_1,\ldots,\Omega_N):$$

$$\bigcup_{i=1}^{N} \Omega_i = \Omega, mes\left(\Omega_i \cap \Omega_j\right) = 0,$$

$$i \neq j, i, j = \overline{1,N}\};$$

- $\tau_i = \left(a_1^i, a_2^i, b^i, \nu_i\right), i = \overline{1,N}$ – parameters of functions $\phi_i(t,x;\tau_i) = a_1^i x_1 + a_2^i x_2 + b^i f_i(\nu_i,t)$ – the right-hand sides of (3), which henceforth will be called "centers" of subsets Ω_i ;

- $x(\cdot;\overline{\omega},\tau)$ – solution of the Cauchy problem (3), (4), corresponding to parameter set $(\overline{\omega},\tau) \in P_N(\Omega) \times R^{4N}$.

It is necessary to determine a partitioning $\overline{\omega} = (\Omega_1,\ldots,\Omega_N) \in P_N(\Omega)$ and unknown coordinates $\tau_i, i = \overline{1,N}$ of the centers of subsets $\Omega_1,\ldots,\Omega_N$ whereby the difference between the experimental and calculated data would reach the minimum value, i.e. the minimum value would reach a certain quality criteria:

$$\mathrm{I}\left(x(\cdot;\overline{\omega},\tau)\right) \to \min_{(\overline{\omega},\tau) \in P_N(\Omega) \times R^{4N}}. \tag{6}$$

The paper proposes such way of establishing the problem composite function (6):

$$\mathrm{I}_1\left(x(\cdot;\overline{\omega},\tau)\right) = \int_0^T \beta_0 \left(x_1(t;\overline{\omega},\tau) - \tilde{x}_1(t)\right)^2 +$$

$$+ \beta_1 \left(x_2(t;\overline{\omega},\tau) - \tilde{x}_2(t)\right)^2 \sum_{k=1}^{K} \delta\left(t - t_k\right) dt, \tag{7}$$

where $x_1(\cdot;\overline{\omega},\tau)$, $x_2(\cdot;\overline{\omega},\tau)$ – solution of the Cauchy problem (3), (4) of the recovered system; $\tilde{x}_1(t)$ – the observed phase variable values; $\tilde{x}_2(t)$ –

approximate values of the phase coordinate $x_2(\cdot)$, calculated according to the observations` results of variable $x_1(\cdot)$ by way of numerical differentiation;

$\beta_0 \geq 0, \beta_1 \geq 0, \beta_0^2 + \beta_1^2 \neq 0$ – parameters that define the term priority in the functional (7).

It is obvious that functional (8) is a special case of the functional (7) for $\beta_0 = 0, \beta_1 = 1$ is the following:

$$\mathrm{I}_2\left(x(\cdot;\overline{\omega},\tau)\right) =$$

$$= \int_0^T \left(\dot{x}_2(t;\overline{\omega},\tau) - \dot{\tilde{x}}_2(t)\right)^2 \sum_{k=1}^{K} \delta(t - t_k) dt , \tag{8}$$

where $\dot{\tilde{x}}_2(t)$ – approximate values of the phase coordinate $x_2(\cdot)$ derivative, calculated according to the observations results of variable $x_1(\cdot)$ by the operation of numerical differentiation.

4 INCORRECTNESS OF THE PROBLEM OF MINIMIZING THE FUNCTIONAL (7) AND THE WAYS TO SOLVE IT

Computational experiments showed that the process of recovering the system (3) right side function, which is piecewise continuous on phase variables, on the basis of solving the problem of minimizing the functional (7), has a number of features:

– if a set of observations involves the switching points, then system identification quality depends essentially on whether the sample of observations contains points from different functioning zones;

– the quality of the dynamic system reconstruction in the case of recovery of unknown subsets centers coordinates $\tau_i, i = \overline{1,N}$ depends on choosing the initial approximations of these parameters;

– the problem is characterized by non-uniqueness of solution, which is overcome by narrowing the set where the unknown parameters $\tau_i, i = \overline{1,N}$ are to be searched.

In addition, the main difficulty that arises during the approach implementation is that the problem of function differentiation given approximately is ill-defined in the uniform metric. One of the approaches to solving the above issue is taking into account the error of defining the function while selecting the grid step. Regularized procedure for numerical differentiation can be obtained on the basis of different approaches (Tihonov 1979). At the

155

core of one of the methods, for example, is a traditional approximation of the differential operator by a finite-difference operator.

At the same time the grid step must be agreed with the error of input data. For instance, when the function value can be evaluated at any point and with any step up to δ, determining of the derivative with a given accuracy ε is possible by choosing step $h(\delta) = c\delta^q$ where c and q – positive constants, $q < 1$.

5 SOLUTION OF THE IDENTIFICATION PROBLEM

Let us describe an approach to the dynamical system (3) identification according to observations based on solving the minimization problem of the functional (7).

Considering characteristic functions $\lambda_1(\cdot), ..., \lambda_N(\cdot)$ of the subsets $\Omega_1, ..., \Omega_N$ from the minimization problem of the functional $I_1\left(x(\cdot; \overline{\omega}, \tau)\right)$ subject to (3), (4) we proceed to equivalent problem of infinite-dimensional programming: to find the vector function $\lambda(\cdot) = \left(\lambda_1(\cdot), ..., \lambda_N(\cdot)\right) \in \Lambda_0$ and the vector $\tau = \left(\tau_1, ..., \tau_N\right)$, $\tau_i = \left(a_1^i, a_2^i, b^i, v_i\right)$, $i = \overline{1, N}$, whereby

$$\overline{I}_1(\lambda(\cdot), \tau) \to \min_{(\lambda(\cdot), \tau) \in \Lambda_0 \times R^{4N}}, \qquad (9)$$

where

$$\overline{I}_1(\lambda(\cdot), \tau) = \int_0^T \sum_{i=1}^N p(\mathrm{x}, \mathrm{t}; \tau) \lambda_i(x_1(t)) \sum_{k=1}^K \delta\left(t - t_k\right) dt,$$

$$p(\mathrm{x}, \mathrm{t}; \tau) = \beta_0\left(x_1(t; \tau) - \tilde{x}_1(t)\right)^2 +$$

$$+ \beta_1\left(x_2(t; \tau) - \tilde{x}_2(t)\right)^2,$$

$$\Lambda_0 = \{\lambda(x_1) = (\lambda_1(x_1), ..., \lambda_N(x_1)):$$

$$\sum_{i=1}^N \lambda_i(x_1) = 1; \ \lambda_i(x_1) = 0 \vee 1 \ almost$$

$$everywhere \ for \ x_1 \in \Omega; \ i = \overline{1, N}\},$$

$$\begin{cases} \dot{x}_1 = x_2 \\ \dot{x}_2 = \sum_{i=1}^N \phi_i(t, x; \tau_i) \lambda_i(x_1) \end{cases}, \qquad (10)$$

$$x_1(0) = x_{10}, \quad x_2(0) = x_{20}. \qquad (11)$$

Let us produce a numerical algorithm for solving the problem (9) – (11) under the assumption that time points at which the value of output variable $x_1(t)$ is observed, are equidistant, i.e. $t_k = kh_t$, $k = \overline{1, K}$, where $h_t = T/K$. The proposed algorithm is based on rewriting problem (9) in the following way:

$$I_3(\tau) \to \min_{\tau \in R^{4N}}, \qquad (12)$$

where

$$I_3(\tau) = \min_{\lambda(\cdot) \in \Lambda_0} \overline{I}_1(\lambda(\cdot), \tau). \qquad (13)$$

To solve the minimization problem of function (12) we will apply the generalized gradient descent method with space dilation in the direction of two consecutive generalized gradients – r-Shor's algorithm. The internal problem (13) – the search for vector function $\lambda(\cdot) \in \Lambda_0$ – will be solved on each step of the numerical integration of the system (10), (11) by the formula (14).

Herewith, the step of numerical integration of the system (10) – (11) must coincide with the step h_t.

$$\lambda_i(x_1(t)) =$$

$$= \begin{cases} 1, if \quad p(x^{(i)}, t; \tau_i) = \min_{p=1, N} p(x^{(p)}, t; \tau_p) \\ 0, otherwise \qquad \qquad i = \overline{1, N}, \end{cases} \qquad (14)$$

Where $x^{(i)} = (x_1(t; \tau_i), x_2(t; \tau_i))$ – the phase coordinates, calculated at right side function of (10) equals $\phi_i(t, x; \tau_i)$.

Algorithm.

Initialization. If we take, that m=0, $\varepsilon > 0$, $\tau_i^m = \left(a_1^{im}, a_2^{im}, b^{im}, v_i^m\right)$, $i = \overline{1, N}$; $H_0 = E$ – identity matrix of dimension $(4N \times 4N)$, then we compute (in the case $\beta_1 \neq 0$) $\tilde{x}_2(t) = \dot{\tilde{x}}_1(t)$ at points $t = t_k$, $k = \overline{1, K}$ from the formulas of numerical differentiation; and we set arbitrary values of the characteristic functions $\lambda_i^m(x_{10}), i = \overline{1, N}$.

1. We solve the Cauchy problem (10) – (11) using one of the Runge – Kutta method. At each step of

the numerical integration of (10), we obtain the values of the characteristic functions $\lambda_i^m(x_1(t))$, $i = \overline{1, N}$ according to the formula (14). For this purpose, every step of the system integration should be made for all N possible assignments of the right-hand side of the differential operator.

2. Now we compute the generalized gradient γ^m of the function $\overline{I}^2(\lambda, \tau)$ over the variables $\tau_i^m = \left(a_1^{im}, a_2^{im}, b^{im}, v_i^m \right)$, $i = \overline{1, N}$, when $\lambda = \lambda^m$.

3. Then we conduct m-th iteration of r-Shor's algorithm by the formula

$$\tau^{(m+1)} = \tau^{(m)} - h_m \frac{H_m \gamma^m}{\sqrt{(H_m \gamma^m, \gamma^m)}},$$

where H_m – space dilation matrix with a coefficient α (it is advisable to take equal to 3) in the direction of the difference between two consecutive generalized gradients, which is recalculated by the formula

$$H_{m+1} = H_m + (1/\alpha^2 - 1) \frac{H_m \zeta^m (\zeta^m)^T H_m}{(H_m \zeta^m, \zeta^m)};$$

$$\zeta^m = \gamma^m - \gamma^{m-1}.$$

If due to rounding off, H_{m+1} ceases to be positive definite, we replace it with identity matrix.

The stepper factor $h_m \geq 0$ is selected from the condition of the objective function minimum in the direction of $d = -H_{m+1} \gamma^m$.

If one of the conditions

$$\left\| \tau^{m+1} - \tau^m \right\| \leq \varepsilon ; \quad \left\| \gamma^m \right\| \leq \varepsilon ;$$

$$\left| \overline{I}_1(\lambda^{m+1}, \tau^{m+1}) - \overline{I}_1(\lambda^m, \tau^m) \right| \leq \varepsilon ,$$

is satisfied, then this is the end of the algorithm. As the optimal solution of problem (9) - (11) we choose $(\lambda^{m+1}, \tau^{m+1})$. Then we otherwise set $m := m + 1$ and go to step 1.

Remark. In order to compute the generalized gradient of the function $\overline{I}_1(\lambda^m, \tau)$, it is necessary along with the Cauchy problem on step 1 to solve the $4N$ Cauchy problem at the perturbed components of the parameters τ vector.

6 RESULTS OF MULTI-ZONE DYNAMICS MODEL IDENTIFICATION

Initially, to determine the feasibility of identifying the multi-zone dynamic systems using the proposed approach, the recovering problems of a piecewise linear right-hand side for **one** differential equation were considered: $\dot{x} = a_{0k} x + a_{1k} t + a_{2k}$, $x \in \Omega_k$, $(\Omega_1, ..., \Omega_N) \in P_N(\Omega)$. The research was conducted in the direction of identifying the impact of the initial data (number of observation points, the initial values of the recovered functions parameters) on the quality of the dynamic system reconstruction. Analysis of the computational experiments results led to the following conclusions:

1) multi-score increase in the observations number does not affect significantly the reconstruction quality (Fig. 1);

2) if there are some switching points in the observations set, then the sample of observations must contain points from different functioning zones, their number depending on the form of the right-hand sides;

3) if the dynamic system is characterized by a sliding mode, then using both functionals (7) and (8) allows to identify this property and recover the right-hand side with satisfying accuracy, which is reflected in Fig. 1;

4) for different initial approximations of recovered parameters one can get their different "optimal" values, although it may not affect the quality of the phase curves recovery (Fig. 2).

Let us present the computational experiments results after the recovery of discontinuous on phase variable right-hand side of the system of **two** differential equations (3). We will present a comparative analysis of using the considered functionals below.

Fig. 3 shows the identification results of the three-band dynamic system (3), which can be described by a piecewise linear right-hand side.

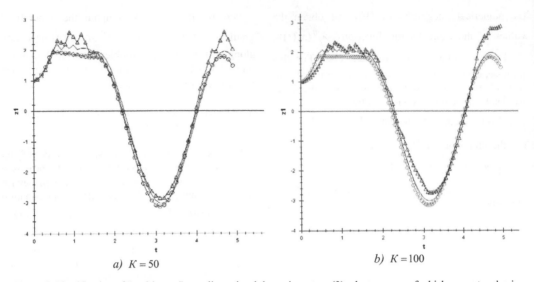

| *a)* $K=50$ | *b)* $K=100$ |

Figure 1. Identification of "multi-zone" one-dimensional dynamic system (3), phase curves of which are restored using approaches based on the solution of the minimization problem of the functional (8) (circles) and functional (7) (triangles) in terms of the results of K observations

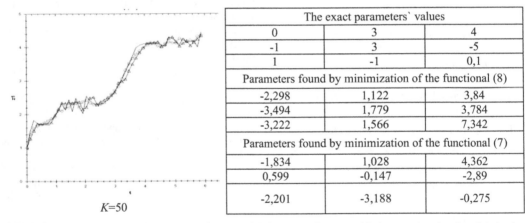

The exact parameters' values		
0	3	4
-1	3	-5
1	-1	0,1
Parameters found by minimization of the functional (8)		
-2,298	1,122	3,84
-3,494	1,779	3,784
-3,222	1,566	7,342
Parameters found by minimization of the functional (7)		
-1,834	1,028	4,362
0,599	-0,147	-2,89
-2,201	-3,188	-0,275

$K=50$

Figure 2. The reconstruction of one-dimensional system with three operating zones (circles correspond to the minimization problem of the functional (8), and triangles - functional (7))

Function $\phi_i = a_{i1}x_1 + a_{i2}x_2 + b_i t$ parameters, the accurate ones and recovered according to the results of 50 observations, are presented in Table 1. As already was noted, such systems are the models of many technical object functioning. They are usually described by nonlinear systems, but can also be presented as a system of linear ordinary differential equations with switchable right-hand sides.

Figure 4 represents the phase curves of the same system, which were recovered according to the results of 200 observations. Analyzing the results of solving this and other problems of dynamic systems reconstruction we can trace:

a) a non-uniqueness of solutions, i.e. parameter values may significantly differ, but the right-hand side approximation is carried out with satisfying accuracy (Table 1). This fact does not contradict the properties of inverse problems;

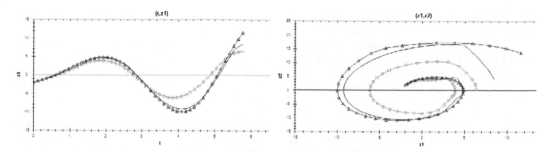

Figure 3. Phase curves of the system (3) ($N = 3$, boundaries between zones $z_1 = 5$, $z_1 = -8$), recovered by minimizing the functional (8) (circles) and functional (7) (triangles) in terms of $K = 50$ observations

Table 1. The parameter values of a piecewise linear function of the equations system (3) right-hand side, the phase curves of which are shown in Fig. 3

i	Exact values of parameters of φ_i	Initial approximation of parameters	Recovered parameter values using an approach based on:	
			Minimizing the residual functional (7)	Minimizing the residual functional (8)
1	-2; 0.4; 0	0.63; 0.44; -0.27	-1.96; 0.65; 0.23	-2.88; 1.89; 1.00
2	0; 1.5; -3.5	0.50; 0.55; 0.79	-1.98; 0.63; 0.32	-1.89; 0.19; 0.28
3	-2.4; -2; 0.3	0.98; 0.38; 0.63	-1.30; 0.78; -1.84	-2.09; -1.50; 0.20

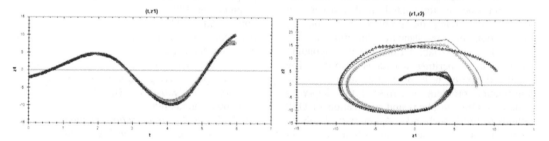

Figure 4. The phase curves of the system (3) ($N = 3$, boundaries between zones $z_1 = 5$, $z_1 = -8$), recovered in terms of $K = 200$ observations

b) an increase in number of observations improves the reconstruction quality of dynamic system (Fig. 3, 4) using an approach based on minimization of the functional (8), which is not always true for the approach on solving the problem of minimization by the functional (7) (Fig. 5).

c) the boundaries between functioning zones of the system are identified with sufficient accuracy (Fig. 3 – 5);

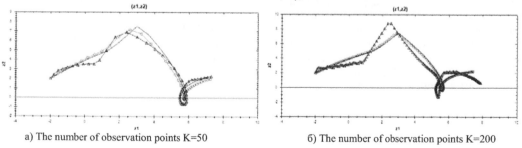

a) The number of observation points K=50

б) The number of observation points K=200

Figure 5. The reconstruction of two dimensional system (3) with three functioning zones

It should also be noticed that the quality of a dynamic system reconstruction depends on the initial approximations of the unknown parameters. This is explained by the fact that for the solution of the optimization problem, the non-differentiable optimization techniques that generally converge to a local minimum are applied.

As stated above, dynamic systems, which can be described by piecewise linear right-hand side, are models of many technical objects functioning that can be described, in general, by nonlinear systems, but can also be quite well presented as a system of linear ordinary differential equations with switched right-hand sides.

7 CONCLUSIONS

The article is a consistent extension of the theory of continuous OSP problems development in the area of working-out and validation of methods for solving dynamic optimal partitioning problems and their practical applications. The problem of dynamic system identification, which is described by a system of differential equations with discontinuous right-hand side, is formulated in this paper. An approach to solving the problem based on the minimization of the standard deviation of the recovered phase variables from the corresponding values of the experimental data is proposed. The considered approach involves formulation of the identification problem in terms of the theory of continuous OSP problems and application of OSP methods to fulfill such tasks. A key part of solving algorithm is a r-Shor`s algorithm of non-differentiable multivariate function minimization. The method and algorithm for the reconstruction of systems of ordinary differential equations with discontinuous right-hand side based on combining elements of the theory of inverse problems for differential equations and continuous OSP problems, as well as the theory of non-smooth optimization problems are developed.

The proposed mathematical apparatus can be used for solving optimal control problems for the set of facility operating states when in the course of control process, the parameters of dynamic model as well as quality functional may change, etc. The solution method and models presented in this paper, can be further generalized to the case:

– of the presence of a delay in phase coordinates or control parameters;
– when the object or process dynamics is presented not by observations of one of phase coordinates, but by a linear combination of values of several phase coordinates. In this case the question about the possibility of dynamic system recovery from the observation data raises;
– of the transition to the optimal control problem as one of the approaches to the solution of dynamic systems identification problems and application of the theory of lumped systems control methods (for instance, indirect, based on the Pontryagin maximum principle);
– of solving the problem of identification in a broad sense involving the theory of operator equations along with the OSP methods.

REFERENCES

Matveykin V.G., Muromtsev D.I. 2007. *Theoretical basis of energy-saving control in dynamic regimes installations of industrial and technical purpose* (in Russian). Moscow: Publisher Engineering: 128.

Filippov A.F. 1985. *Differential equations with discontinuous right-hand side* (in Russian). Moscow: Nauka: 225.

Kiseleva E.M., Koriashkina L.S. 2013. *Models and methods for solving continuous problems of optimal sets partition* (in Russian). Kiev: Naukova Dumka: 606.

Shor N.Z. 1979. *Minimization methods of non-differentiable functions and these applications* (in Russian). Kiev: Naukova Dumka: 199.

Bessonov A.N., Zagashvili I.V., Markelov A.S. 1989. *Methods and means of dynamic objects identification* (in Russian). Moscow: Energoatomizdat: 280.

Grop D. 1979. *Methods of systems identification* (in Russian). Moscow: Mir: 472.

Tikhonov A.N., Arsenin V.I. 1979. *Methods for solving ill-posed problems* (in Russian). Moscow: Nauka: 288.

Vatulyan A.O. 2007. *Inverse problems in solid mechanics* (in Russian). Moscow: Fizmatlit: 223.

Isermann R., Münchhof M. 2010. *Identification of Dynamic Systems: An Introduction with Applications*. Springer: 730.

Power Engineering, Control and Information Technologies in Geotechnical Systems – Pivnyak, Beshta & Alekseyev (eds)
© 2015 Taylor & Francis Group, London, ISBN 978-1-138-02804-3

Identification, prediction and control of complex multiply technological objects

V. Kornienko, I. Gulina & Yu. Rybalchenko
State Higher Educational Institution «National Mining University», Dnipropetrovs'k, Ukraine

ABSTRACT: Identification of control object is executed. Characteristics of a condition of generating process are certain. Accuracy of intellectual prognosis of parameters of complex multicoherent object is estimated. On an example of modelling of an adaptive control system blast-furnace process estimates quality of control at use of self-tuning regulators.

1 INTRODUCTION

For control of complex controlled objects (CO) characterizing by nonstationarity, nonlinearity and stochasticity, it is actual the identification and prediction problem solving, which enables to improve the quality of control by such CO at the cost of rising of evaluation precision of its condition.

2 FORMULATING THE PROBLEM

From the position of control, the compound CO is, for example, blast-furnace process, the control of which presupposes the ensuring of the predetermined chemical composition and temperature of the hot metal at the output stage that is defined by the terminal state (TS) of the blast furnace in the process of iron smelting.

One of the most unbiassed estimation method of BF TS is the control of cycles of 'accumulation-emission' of the liquid products of melting, characterizing by the change of the level of fusion (LF) (Gusev, Makienko & Rybalchenko 2010). Though, in view of oscillatory character of that signal, the interest for TS estimation is not in instantaneous values, but in LF trend.

The uncertainty of condition of the compound CO makes necessary to use the adaptive automatic control system (CS), in which it is executed the adaptation of control models and laws to the real operating conditions of CO.

For identification and prediction of CO with uncertainty it is used the adaptive filters-approximators (AFA) (Gulina & Kornienko 2011), among which the most prospective are the AFA on the basis of methods of intelligence systems, particularly, neuronetwork (NN) and fuzzy systems,

so far as they are universal and effective approximators (Kruglov, Dli & Golunov 2001).

Thereby, the research of means of identification, prediction and control of compound CO is an open problem.

3 IDENTIFICATION

The purpose of identification is the acquisition of the CO mathematical model, suitable for use in the control loop. It is carried out on the basis of experimental data about CO.

For choice of Δt it can be used the condition of transition from continuous interval to its discrete representation, as well as the condition of ensuring of needed prediction depth (for compensation of delay in a system).

It is known (Gusev, Makienko & Rybalchenko 2010) that on the channel 'ore burden – content of silicon in the cast iron' (top control of BF), the time constant is T = 2,6...6 h, and the delay is τ = 4...4,5 h. And on the channel 'rate of blowing – content of silicon in the cast iron' (bad control of BF), the time constant is T = 1...5 h, and the delay is τ = 0,5...1 h.

In the condition of BF-6 of Nizhnetagilskii Metallurgical Works, the average time between the outputs of the cast iron (melt duration) is $\bar{t}_u = 155$ min.

The value of control interval Δt (and discretization) is determined as by dynamic properties of CO, so by spectral characteristics of its signals.

For choice of Δt it could be make the system of equations:

$$\Delta t = \bar{t}_h / N_{app} ; \tag{1}$$

$$\Delta t = T_{r\min} / N_{app} ; \qquad (2)$$

$$\Delta t + \tau_{\max} \leq T_{pr} \leq \tau_{cor} , \qquad (3)$$

where N_{app} – the number of points of the time interval approximation ($N_{app} = 5...10$); $T_{r\min} = m \cdot T_{\min}$ – minimum time of regulation ($m = 3...5$), and $T_{\min} = 1$ h $= 60$ min – minimum value of the time constant of the control channels of BF TS from the top and bottom); T_{pr} – the needed depth of prediction; $\tau_{\max}$ – maximum delay on control channels ($\tau_{\max} = 4,5$ h $= 270$ min); τ_{cor} – correlation window of the controlled value signal.

The expressions (1)–(2) define the condition (according to the Wedrow theory) of transition from one continuous interval to its discrete representation, and the expression (3) – the condition of ensuring of the needed depth of prediction T_{pr}, which, in general case, cannot be more than τ_{cor} (Kornienko & Skryl' 2009).

Then, according to (1) we shall get $\Delta t = 15...30$ min, and according to (2) – $\Delta t = 12...60$ min, whence beforehand we can get the value $\Delta t = 30$ min.

For definition of dynamic properties of signal, characterizing the LF, it was built the time-frequency representations under its instantaneous values (Fig. 1).

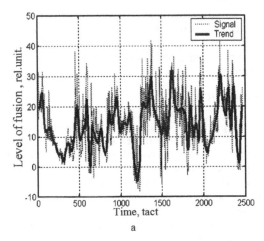

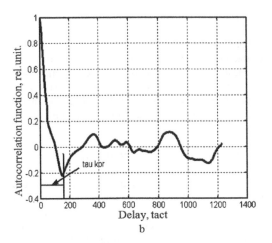

a   b

Figure 1. Signal of LF (Signal) and its trend (a), and autocorrelation function of trend (b)

Here the LF trend is received by averaging on value of the melting time $\bar{t}_h$, using the wavelet function in the form of symlet (Dremin, Ivanov & Nechitaylo 2001; Kornienko, Kuznetsov & Garnak 2009).

Having specified the correlation confidence level 0,2 according to Fig. 1,b we get the correlation interval for LF trend $\tau_{cor} = 158 \cdot 5$ min $= 790$ min. Thus, the condition (3) is satisfied for adopted $\Delta t = 30$ min , since 30 min + 270 min < 790 min.

For compensation of maximum delay on control channels, the depth of prediction n has to make (see (3)):

$$n = (\Delta t + \tau_{\max}) / \Delta t = 10 \text{ tacts.} \qquad (4)$$

For correct identification it is preliminarily needed to sort out the useful signal from the noise.

One of the effective smoothing (averaging) method of signals is its low-frequency filtration, realizing, for example, with the help of lowpass filters (LPF).

For receiving of LF trend in such LPF, the cutoff frequency is $f_{cut} = \bar{t}_h^{-1}$, and the discretization period of signal shall not be more than the control interval of BF LF Δt.

When realizing such LPF, it should be remembered that the temporal filters have the essential phase shift (Gulina & Kornienko 2011). At the same time, the filters with frequency or time-frequency (wavelet) transformations have not that imperfection.

Since the signal of BF LF is nonstationary, for its selection it is needed to choose the time-frequency LPF on the wavelet basis with the properties of orthonormality, compactness and restoration without loss (Dremin, Ivanov & Nechitaylo 2001).

162

The scheme of that filter is showed below on the Fig. 2.

Figure 2. The scheme of wavelet LPF

Here the filtration is based on digital wavelet transformer (DWT) of the signal, threshold restriction of the coefficients of wavelet expansion – tresholding (Tresh) and return DWT (RDWT) of the modified wavelet coefficients.

Wavelet expansion of signal of LF $U[k]$ is carried out on several levels of specification: $l = \overline{1, L}$. At tresholding it is necessary to throw off the wavelet coefficients of specification of the expansion levels l, having the scale m_l, less than average time between cast iron output $m_l < \bar{t}_h$.

Then the ODWT on wavelet coefficients of approximation and modified coefficients of specification form the LF trend $\hat{U}[k]$.

Since $\bar{t}_h / \Delta t \approx 5$, so the number of expansion levels of wavelet coefficients of specification is accepted as $L = 3$. At that time, the coefficients of expansion levels 1 and 2: $2^1 = 2 < 5$ and $2^2 = 4 < 5$ are zeroed, and the coefficients of the level 3: $2^3 = 8 > 5$ are limited.

In the capacity of basis wavelet it was used the symlet of the fifth degree.

The wavelet coefficients of signal of LF and its trend are showed on the Fig. 3, and results of filtration of wavelet signal of BF LF are showed on the Fig. 4.

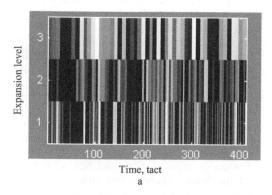

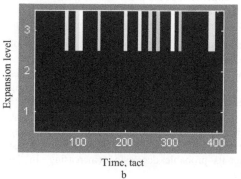

Figure 3. Wavelet coefficients of LF signal (a) and its trend (b)

On that figure it is seen that the delay between the signal and its trend is absent. Besides, the wavelet LPF has better smoothing properties as compared with linear digital LPF that is obviously from the Fig. 5.

Let's define the characteristics of the condition of the process, causing the BF LF trend.

The calculations allowed to determine the correlation entropy K_s, correlation dimension D_s and Hearst value H for LF trend.

Its values are: $K_s = 0,49$, $D_s = 2,21$ and $H = 0,21$. Herewith, the interval of precision predictability (depth of precision prediction) is $T_s = 6,17$ tacts (with duration Δt). For more time it is possible use only statistical prediction, the depth of which depends on autocorrelation function of the process.

For determination of the dimension d of the phase space (memory depth) of LF trend it was calculated its upper estimate $d \leq 5$.

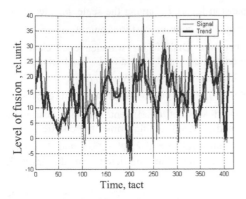

Figure 4. LF signal and its trend,
received with the help of wavelet LPF

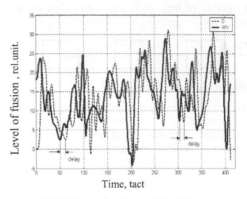

Figure 5. BF LF trend on linear digital (D)
and wavelet (WV) LPF

The dimension of d is the dimension of attractor placement (condition of generating system), i.e. the whole dimension of the phase space, which includes the whole attractor.

The correlation dimension D_s practically stops to increase (come into satiation) at the dimension of phase space of $d \geq 3$. Thus, $3 \leq d \leq 5$.

The Hearst value allows to classify the time series and to determine the evolution of its generating systems. Its value $H = 0,21 < 0,5$ characterizes the LF trend as the ergodic changeable process, consisting of frequent falls and rises. I.e. increasing (decreasing) of the values of the process in the past means its probable decreasing (increasing) in the future.

According to the methods of identification of nonlinear dynamic processes (Kornienko & Skryl' 2009), the second stage of identification is the reconstruction of the model of process, generating the BF LF trend.

At the same time, it is made the choice of the type of predictive AFA (with its basis functions and parameters), which are adjusted under experimental signals in optimal way according to the chosen fidelity criterion.

4 PREDICTION

For prediction of LF trend it was used the linear (Lin), neuro fuzzy (Anfis), neuronetwork (NN) and neuro wavelet (WVNN) AFA (Fig. 6).

In the capacity of Lin AFA it was used the adaptive FIR-filter of 8th degree with normalized gradient algorithm and adaptation tact equal to 0,75.

In WVNN AFA (Kornienko, Kuznetsov & Garnak 2009) it was used the wavelets of symlet type of the 5th degree with two expansion levels and cascade NN with 32 neurons in the buried level with sigma functions of activation and 1 linear neuron in the output layer.

For construction of Anfis and NN AFA it was used the Wiener-Hammerstain structure with the number of delay lines at the input (memory depth) equal to dimension of phase space of LF trend $d = 4$, determined above.

In the capacity of Anfis AFA it was used the system of the fuzzy output of Sugeno structure (Kruglov, Dli & Golunov 2001) with bell-shaped membership function and 15 epochs of training.

In NN AFA it was used the cascade NN of direct distribution with 64 neurons in the buried layer with sigma functions of activation and 1 linear neuron in the output layer, and the number of training cycles limited to 300.

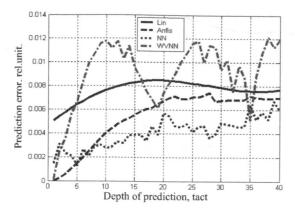

Figure 6. Errors of prediction of LF trend for different AFA: linear (Lin), neuro fuzzy (Anfis), neuronetwork (NN) and neuro wavelet (WVNN)

In the capacity of Anfis AFA it was used the system of the fuzzy output of Sugeno structure (Kruglov, Dli & Golunov 2001) with bell-shaped membership function and 15 epochs of training.

In NN AFA it was used the cascade NN of direct distribution with 64 neurons in the buried layer with sigma functions of activation and 1 linear neuron in the output layer, and the number of training cycles limited to 300.

The original sequence of LF trend was divided into taught and checking sequences equally, and in the capacity of optimization criterion it was used the relative mean square error between real and predicted value of LF trend in checking sequence.

The modeling of AFA was carried out with the prediction depth of 40 tacts ($40 \cdot \Delta t$).

The analysis of received errors (see Fig. 6) shows that the error of prognosis doesn't exceed 1,2 %. At the same time, the least errors have NN AFA.

In the same way it was carried out the estimation of prediction errors of the indices of cast iron quality at the stage of emission (Fig. 7). The prediction depth is 8 outputs ($8 \cdot \bar{t}_h$). At the fig. 7 there are the graphics of errors of the AFA, which gave the best result.

When modeling, all indices were normalized within the limit ± 1, and its average values were: cast iron temperature – 1481 °C, content of silicon – 0,755 % and content of sulphur – 0,022 %.

It is determined that the relative error of prediction for cast iron temperature doesn't exceed 1 %, and for the content of silicon and sulphur – 5 %.

The statistical check on nonparametric criterion of signs confirmed that the results of prediction with

the significance level 0,01 are adequate to experimental realizations.

5 CONTROL

Adaptive CS allow to reduce the terms of design, adjustment and test, as well as to ensure the effective control in disturbance conditions and change of CO properties.

The preferable for CS of BF LF is the realization of the principle of searchless indirect adaptive control (Gulina & Kornienko 2011), since the major loop is opened, that causes the CS asymptotic stability.

In searchless adaptive CS preliminarily it is carried out the CO identification, and then, knowing the CO parameters, the coefficients of regulators are calculated.

Since BF LF has the significant delays on control channels, so it is used CO prediction models realizing with AFA help for its compensation.

Let the CO dynamics is described by nonlinear difference equation:

$$\hat{x}[k+n] = F\{x[k],u[k],w[k],\xi[k],a[k],k\};$$

$$k = \overline{k_j, k_{j+1}-1}, \quad j = 0,1,2\ldots, \qquad (5)$$

where F – aggregate transfer function; $\hat{x}[k],u[k],w[k],\xi[k],a[k]$ – vectors (matrices) of CO state estimate, its control, disturbance, noises and parameters up to the current time k ; k_j,k_{j+1} – initial tacts of sequential control cycles; n – needed prediction depth.

165

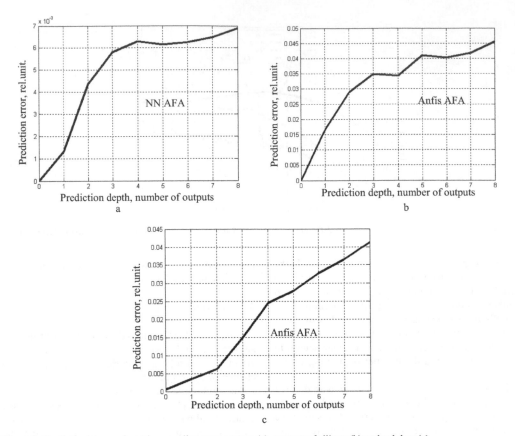

Figure 7. Prediction errors of cast iron quality: temperature (a), content of silicon (b) and sulphur (c)

The offered CS (Fig. 8) consists of local subsystems (loops 1 and 2). In it according to the setting valuation $q[k]$ the regulators REG1 and REG2, parameters of which are corrected by results of CO identification with the help of adaptation of coefficients $\hat{a}_1[k], \hat{a}_2[k]$ AFA1 and AFA2, produce the control actions $u_1[k]$ and $u_2[k]$, affecting the CO state $x[k]$.

As a rule (Gulina & Kornienko 2011), the adaptive CS shall be constructed as the optimal as for control quality that is expressed in fidelity of reproduction of master control q:

$$J = \|q[k] - \hat{x}[k]\| \to min .$$ (6)

For CO identification it founded the distribution the searchless algorithms of parametric identification with adaptive model, oriented on functioning in the real time scale, which are, for example, gradient algorithms. In that case the identification process consists in adaptation of parameters a on the value of error functional J_e between the real output and model response (functional gradient in parameter space):

$$\hat{a}[k] = \hat{a}[k-1] - K_a \cdot \nabla_{\hat{a}} J_e\{x[k], \hat{x}[k], k\}$$ (7)

where $\hat{a}[k]$ – estimate of vector of the adaptable parameters in current tact; $\nabla_{\hat{a}} = (\partial / \partial \hat{a})^T$ – symbol of gradient; K_a – matrix of coefficients.

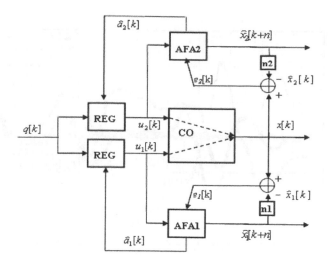

Figure 8. The structure of CS by multiply CO

The error functional J_e has the following form:

$$J_e = \frac{1}{2} E\{(e[k])^2\} = \frac{1}{2} E\{(x[k] - \hat{x}[k])^2\} \qquad (8)$$

where E – expectation value, $e[k] = x[k] - \hat{x}[k]$ – error.

AFA (CO model) becomes optimal at $\hat{a} = a_{opt}$, when $J_e = 0$, then the adaptation problem consists in finding of optimal coefficients by the way of iterative determination of the gradient of surface of minimum mean square error (8).

Since AFA use NN and neuro fuzzy systems, carry out the vector data processing, so the accounting of disturbances $w[k]$ in the expression (5) doesn't cause difficulties of principles (its accounting increases the dimension of the AFA input data, not changing the processing algorithm. In connection with that, then the accounting of disturbances is not considered at the control synthesis.

At realization of the adaptive CS it can be used the regulators (REG1 and REG2 on the Fig. 8) of different types, as self-tuning, so standard (PI, PID).

The estimate of quality of adaptive CS is made by the way of simulation on the basis of experimental data received in conditions of BF-3 of Mariupol Metallurgical Works named after Ilich.

The models of control channels on the top and on the bottom were represented as aperiodic links with delay. from the peculiarities of CO the prediction depth on control channel on the top is equal to 10 tacts (3,5 hours), and on the control channel on the bottom – 3 tacts (1 hour). At the same time, the memory depth according to results received above is equal to $d = 4$ tacts. As a set-point it was used the zero normalized value of LF trend, and the observation period was 256 tacts (approximately 3,5 days at the chosen control interval). The nonstationarity of CO was 10 % from the nominal values of its parameters during the observation period.

For prediction of CO state it was used the NN AFA with cascade NN of direct distribution with 64 neurons in the buried layer and sigma functions of activation and 1 linear neuron in the output layer, and the number of training cycles was limited to 300.

It was simulating the work of adaptive CS with the self-tuning controllers and predictive model (Gulina & Kornienko 2011). At that time, it was used the method of golden section for search of control action optimal by accuracy.

The relative mean square errors were used as the measure of accuracy of prediction and control.

The results of CO control (Fig. 9) include the value of LF trend in uncontrolled (Plant nContr) and controlled (Plant wContr) regimes, outputs of loops of AFA top out and AFA bot out, as well as the changes of control actions on the top (Control top) and on the bottom (Control bot).

167

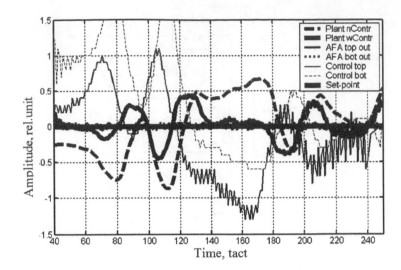

Figure 9. Results of BF TS control on LF trend

The analysis of received results shows that the decrease of the mean square deviation of LF trend from the predetermined value (control error) in the controlled mode relative to uncontrolled one is in 1,59 times.

The statistical check on sign criterion confirmed the significance of the received conclusions.

In the whole, the control errors are determined by the prediction errors. The pre-training of AFA improves the convergence of adaptation algorithm, and accordingly, ensures the decrease of control error.

The time of control synthesis per one tact of prediction at calculation on the processor Pentium IV is approximately 1.1 s, that doesn't make time limitations when using these systems in the control loop of BF TS.

6 CONCLUSIONS

It is grounded the significance of the control interval of BF TS, which takes into account the dynamical parameters of the control channels and spectral properties of CO signals. This allowed to identify the indices of the LF trend characterizing the BF TS on the basis of methods of nonlinear dynamics.

It was developed the linear digital and wavelet lowpass filters, allowing to estimate the trend of melt level of the blast furnace, which characterizes its thermal state.

Under the experimental data it was carried out the identification of BF TS that allowed to estimate the prediction precision with the help of intelligent tools

of LF trend and indices of the cast iron quality at the output.

It was developed the adaptive CS with intelligent prediction of CO state using the local adaptive CS with self-tuning and PID regulators, which ensure the qualitative control of the compound multilinked CO with the disturbing environment of functioning.

By modeling it was ascertained that the use of adaptive CS of BF TS allow to decrease the mean square error of control in 1,59 times. At the same time, the costs for control synthesis don't make the time limitations for using these systems in the control loops of BF TS.

The subsequent researches shall be aimed at creation of automated controlling system of multilinked CO.

REFERENCES

Gusev A.Yu., Makienko V.G. & Rybalchenko Yu.P. 2010. *Evaluation and prediction of the blast furnace stroke through a modified Kalman filter* (in Russian). Dnipropetrovs'k: Scientific Bulletin of National Mining University, Issue 2.

Gulina I.G. & Kornienko V.I. 2011. *Identification and prediction of the blast furnace thermal state at the level of the melt* (in Russian). Mining electrical engineering and automation, Issue 87.

Kruglov V.V., Dli M.I. & Golunov P.Yu. 2001. *Fuzzy logic and artificial neural networks* (in Russian). Moscow: Fizmatlit.

Kornienko V.I. & Skryl' D.Yu. 2009. *Dentification of nonlinear processes in time implementations* (in Ukrainian). Dnipropetrovs'k: Scientific Bulletin of National Mining University, Issue 3.

Dremin I.M., Ivanov O.V. & Nechitaylo V.A. 2001. *Wavelets and their uses* (in Russian). Successes of physical sciences, Vol. 171, Issue 5.

Kornienko V.I., Kuznetsov G.V. & Garnak I.V. 2009. *Wavelet neural prediction and identification of complex signals and control objects* (in Ukrainian). Information Technology and Computer Engineering, Issue 2 (15).

Gulina I.G. & Kornienko V.I. 2011. *Adaptive ACS complex multiply object management with intelligent forecasting* (in Russian). Information processing system, Issue 8(98).

Power Engineering, Control and Information Technologies in Geotechnical Systems – Pivnyak, Beshta & Alekseyev (eds)
© 2015 Taylor & Francis Group, London, ISBN 978-1-138-02804-3

Simulation of methane concentration process control at coalmine

O. Aleksieiev & T. Vysotskaya
State Higher Educational Institution "National Mining University", Dnipropetrovs'k, Ukraine

ABSTRACT: The subject under the issue will always be topical, as it is impossible to create a coal mine methane monitoring and control system which is perfectly safe. The article suggests a linear variant of methane concentration smooth regulation.

1 INTRODUCTION

Coal mines in Ukraine, compared to other countries mines, are characterized by complex mining and geological conditions. The average depth of bank excavation is more than 800 m. 33 mines operate at the depth of $1000 \div 1600$ m. 90 % of 190 operating mines are methane concentration hazardous, 60% are sudden coal dust outbursts prone, 45% are prone to sudden outbursts and rock bumps, 22% are spontaneous coal combustion prone. Ukraine takes the tenth rank for coal production and one of the front ranks for the number of fatal accidents at mines. Notwithstanding coal production and the number of miners considerable curtailment during the last 15 years, the level of mine accidents and fatal injuries rate is still high.

2 MAIN RESULTS PRESENTATION

A mining section with a single longwall constitutes the technological object of the research. The task of the control is to secure methane concentration desired value in the return air of the ventilation scheme: C_{des}, within the limits of (C*perm. lower $\div$ C*perm. upper). In the process of regulation the concentration of methane must not exceed the limits of C_{perm} maximum permissible value, i.e. 1%. The decrease of methane concentration value – compared to C_{des} – should also be worked through with the regard to control economic feasibility, as the reduction of the mine working ventilation air consumption drastically reduces energy costs. Varying the values of air consumption in the system we regulate methane concentration. These values of air consumption are defined by the values of aerodynamic resistance and depression.

Thus, the task of the control is the process of the ventilation scheme parameters change with the aim of methane concentration value reduction to C_{des}, not exceeding the limits of maximum permissible value C_{perm}.

The simplest way to solve the problem of regulation is the straightforward change of the air consumption rate from Q_0 to Q_{des}, where Q_0 is the air consumption rate at the current time and Q_{des} is the air consumption rate securing methane concentration value C_{des}.

According to the investigations (Abramov, Tyan 1973) we can conclude that drastic increase of air consumption may cause methane concentration changes accompanied by lasting transient processes with the concentrations considerably exceeding C_{perm}.

Aforesaid suggests the inadmissibility of drastic air consumption increase in the process of methane concentration regulation.

Linear, safe regulation allows to avoid methane concentration permissible value exceeding and to get considerable reduction of the transient process duration.

The essence of the method lies in the linear increase of the air consumption from Q_0 to Q_1 (here, $Q_1 \geq Q_{des}$) and in the subsequent linear decrease of the consumption to Q_{des}. The increase is done during the time period $dt1$; the decrease is done during the time period $dt2$. Air consumption linear decrease allows to reduce the time of methane concentration outburst attenuation.

The disadvantages of this algorithm are as follows: relative complexity of time intervals $dt1$ and $dt2$ calculation (theses intervals provide the shortest safe transient process and small acceleration of transient processes in the mine ventilation system).

The concerned complex object of control is characterized by the following peculiarities:

1.   Not all targets of controlling solutions and conditions choice, influencing this choice, can be expressed in a form of quantitative ratios.

2.   Control object formalized description is either lacking or extremely complicated.

3.   A significant part of information necessary for the object mathematical description exists in the form of the natural language sentences: they are the ideas and the suggestions of specialists and experts who work with this object.

In the case of complex objects control a person-operator controlling the object, not the object itself, is simulated. Certainly, only a qualified, experienced in hand operation is worth simulating.

In our case we consider a panel with a single longwall as a complex object.

The analysis of methane concentration value and change rate constitutes the essence of the algorithm. Its aim is to predict probable positive and negative bursts and to form necessary compensating changes of the air consumption. Forming a current increase of the air consumption we estimate the proximity of current methane concentration to the maximum permissible value, preventing its exceeding.

One of the Donbass mines is chosen for the investigation of the proposed methane concentration control method. The mine is a typical one for this region. Its output is 1000 tons of coal per day at the average.

Let us consider this mine longwall with the highest gas concentration (Fig. 2) – the 3-d East longwall of the central panel H 4 N seam (Fig. 2). Its geometrical parameters are as follows:

-       wide place length – 198 m;
-       seam height – 0,9 m.

Let us describe the mode of the control object.

Panel mode dynamics according to the factor: air consumption – methane discharge is described in the following form (Boyko & Petrechuk 1972).

$$W_y(P) = (K_{ecf} \frac{T_{gtc}P-1}{T_{gtc}P+1} - \frac{K_l}{T_lP+1}) \frac{K_{lcf}\ell^{-\tau P}}{T_lP+1},$$

where

F_{lcf} - longwall carryover factor, $F_{lcf} = \dfrac{C_0}{Q_0}$;

T_{gtc} - goaf time constant;
K_{ecf} - excavation carryover factor;
T_l - longwall time constant;

F_{lcf} - longwall carryover factor.

According to the task of the investigation, the simulation of the aeration object must meet the following requirement: the adequacy of the information transformation in the simulation and in the real object, i.e. input and output processes qualitative and quantitative correlations obtained in the simulation must be the same as in the real object.

Output methane concentration can be written down as (Abramov, Tyan 1973)

$$C(t) = \frac{q(t_p) + q_k(t_p) + q_c(t_p)}{Q(t_p) + Q_k(t_p) + Q_c(t_p)}, \quad (1)$$

where $q(t_p), Q(t_p)$ - is the mathematical expectation of methane and air discharge respectively; $q_k(t_p), Q_k(t_p)$ - are quasidetermined methane and air discharge constituents; $q_c(t_p), Q_c(t_p)$ - are random methane and air discharge constituents.

In a static regime the functional connection of methane discharge and air consumption is linear. In a dynamic regime under step-like air consumption growth a transitive process of methane discharge is developed (Boyko & Petrechuk 1972)

$$h_q(t_k) = 0,01\{[at_k^m e^{\alpha t_k} + b(1-e^{\gamma t_k})]Q^*(t_k) + C(t_{k-1}), \quad (2)$$

where $C(t_{k-1}), Q^*(t_k)$ - are the values of methane concentration and air discharge at the moments of time t_k and t_{k-1} respectively; a, m, α, b, γ - are constant factors determined on the transitive process curve, the process is stipulated by the air discharge unit step (1 m3/min); a, m, α, b, γ - the time from the previous gas-dynamic process.

Safety door (SD) secures the consumption of air entering the longwall through the airway control (Boyko & Petrechuk 1972). Fig. 1 presents the characteristic of this controller. The repositioning of the safety door leafs is remotely controlled (from the control center). The information on the quantity of air entering the longwall and methane concentration in the longwall atmosphere is transmitted to the surface.

Technological processes static characteristics and pre-determined measurement accuracy determine the period of mine atmosphere state inspection sensors scanning. According to the results of the investigations (Abramov, Tyan 1973) it is found that on the criterion of accuracy the interval of methane concentration inspection sensors scanning must be around 20÷30 min. This period of time also determines the clock cycle of fresh air entering the longwall discharge control, as there are possible significant changes of methane concentration in the goaf during this period.

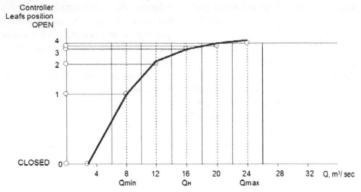

Figure 1. SD controller flow rate curve.

Now, let us look into the process of air consumption control. This control should be done so that not to cause drastic increase of methane concentration under saltatory variation of the controller leafs position change. The point is that SD construction makes it possible to perform a smooth stepless change of the regulator leafs from one fixed position into another necessary one.

Securing smooth variation of the air discharge, it provides necessary smooth variation of methane concentration in the longwall.

On the basis of the above stated command and controlling parameters mutual interaction conditions it is possible to draw a scheme of the air consumption influence on methane concentration (Fig.2).

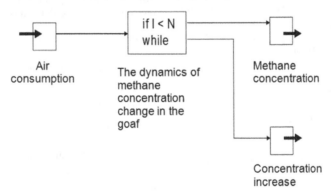

Figure 2. Parameters mutual interaction.

In our case the input variables of the fuzzy logics based controller are as follows:
-methane concentration;
-methane concentration increase.

Methane concentration variable terms: 1) – higher C^*_{perm} - upper; 2) – less C^*_{perm} - upper; 3) – higher C^*_{perm} - lower; 4) – less C^*_{perm} - lower.

Methane concentration increase variable terms:
- positive;
- negative.

Air consumption variation output variable terms:

- small positive;
- considerable positive;
- small negative;
- considerable negative.

It means to change the position of the controller command tools by the value securing necessary changes of the air consumption for the longwall under the question.

The control unit will generate the variable with four terms:
- to open slightly;
- to open considerably;

173

- to close slightly;
- to close considerably.

All terms here are certain intervals of input and output parameters variations. In this case the compliance of the measured values with set parameters is determined on the basis of the membership function in the form of rectangular triangles. Fig. 3 presents set intervals and the membership grade.

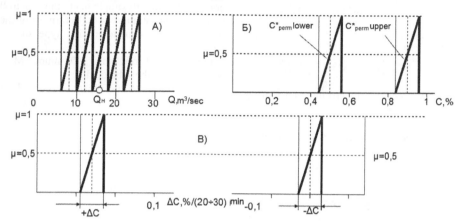

Figure 3. Control set intervals and triangle membership grades.

Gauss curve can be used to describe our terms grades of membership

$$\mu(X) = \ell^{\frac{-(X-c)^2}{2 \bullet sigma^2}},\qquad (3)$$

where $sigma$ - is X variable mean-squared departure;

C – Gauss curve centre.

Mass centre of the membership resulting functions is determined by a well known formula

$$Y = \frac{\int\limits_{Min}^{Max} x\mu(x)dx}{\int\limits_{Min}^{Max} \mu(x)dx}\quad ;\qquad (4)$$

But, for the research of this process, membership functions can also be presented in the form of the triangle. It will by no means adversely affect the accuracy of the simulation; instead, it will considerably simplify the process of control actions choice.

Membership functions for input and output variables are determined by the following formula

$$\mu(x) = 1 - \frac{B-x}{B-A}, x \le B,\qquad (5)$$

where A and B – are upper and lower boundaries of set intervals respectively.

To investigate the process of input and output variables phase correction we will use the Gauss curve (its left part).

Let us present the simplest algorithm of this process. All variables are divided into terms. To do this, we must be aware of:

(the number of terms)
(input variable minimum and maximum value)
(Gauss curve core width)
(if sigma = 0, then Sig = (Maximum – Minimum)/(2*N))

Input variables terms phase correction is done by the following algorithm.

```
begin
input x;
(membership function Gauss curve type)
Function GausFM(x,sigma,c)
GausFM = exp(-(0,5*(x-c)^2)/sigma^2);
End;
(parameters variable arrays for the terms)
Var sigma[TermCount], c[TermCount];
initialization
(the assigning of arrays values)
(input variable minimum value)
    MaxX = Maximum;
(input variable maximim value)
    MinX = Minimum;
(linguistic variable terms number)
    n = TermCount;
    for (I = 1,N)
    begin
    c[i] = MinX+(i-1)*(MaxX -MinX)/(N-1);
    if (Sig <> 0) then
    sigma [i] = Sig else
    sigma [i] = (MaxX-MinX)/(2*N);
```

174

```
      end;
end;
(output membership function array
calculation)
for (i = 1,N)
begin
y [i] = Gaus FM (x, sigma[i], c[i]);
end;
(variable derating control)
if x < MinX then y [i] = 1;
if x > MaxX then y [N] = 1;
output y [TermCount];
end;
```

It is necessary to clarify the last three lines of the algorithm. If the input value equals the range boundary, then according to the algorithm the value of the extreme terms membership function makes

one. If the input value derates, the value of the function should grow smaller; the Gauss curve declines. All these may cause incorrect control decisions making.

Having created the submodel of the phase correction unit to implement the controller of methane concentration in the return air of the goaf, it is possible to simulate different types of the Gauss curve.

Parameters TermCount, Minimum, Maximum, Sig are used in the submodel.

Let us return to the object of control. We are going to control methane concentration in the goaf. Fig. 4 presents the block-scheme of the control.

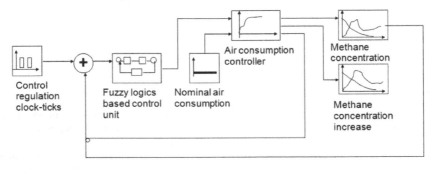

Figure 4. Control block-scheme.

Compositional rules of the output look as follows (Kruglov, Dli & Golunov 2000):

$$R_N \mu_{Q_i}(q) = \max\min[\mu_{Q_i}(q) \max\min[\mu_{C_j}(C) \min \mu_{\Delta C_k}(C, \Delta C)]],$$

$q \in Q_i \qquad c \in C_j \qquad \Delta c \in \Delta C_k$

$c \in C_j \qquad \Delta c \in \Delta C_k$

$\Delta c \in \Delta C_k$

where $\mu_{Q_i}(q)$, $\mu_{C_j}(C)$, $\mu_{\Delta C_k}(C, \Delta C)$ - are the functions of $q, C, \Delta C$ membership in set intervals.

The rule base for the control:

1)   if methane concentration is higher C^*_{perm}-upper, methane concentration increase is negative, then air consumption variation is small positive;

2)   if methane concentration is higher C^*_{perm}-upper, methane concentration increase is positive, then air consumption variation is considerable positive;

3)   if methane concentration is higher C^*_{perm}-lower, methane concentration increase is positive, then air consumption variation is small positive;

4)   if methane concentration is higher C^*_{perm}-lower, methane concentration increase is negative, then air consumption variation is small negative;

5)   if methane concentration is lower C^*_{perm}-lower, methane concentration increase is negative, then air consumption variation is considerable negative.

As we deal with fuzzy logics, it is necessary not just to get a yes-no answer to each rule, but to calculate the degree of its validity.

We have already obtained the entire datum necessary for the creation of fuzzy logics based control unit. Fig. 5 presents its inner structure.

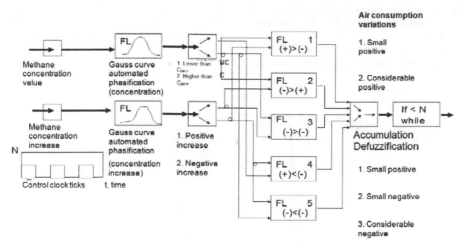

Figure 5. Control unit inner structure.

Accumulation and phasing processes.

Having obtained the state of five deductions from the rules of fuzzy logics, it is necessary to calculate the value of the output variable (that of the air consumption). We shall do it by Mamdani algorithm.

The second part of the program executes the calculated membership functions formation.

Fig. 8 presents this membership functions.

The next part of the listing is given below. The arrays of variable triangle functions of the output variable are described here. In initialization section the assigning of values to the variables for each term is done. As a matter of check-up convenience, let us assign the values to the input vector action.

(The parameters for each term of the output variable)
```
var A[10], B[14],X;
    A[6], B[10], X;
    A[18],B[22], X;
    A[22],B[26], X;
initialization
        (N - the number of output terms)
const N = 4;
(Arrays values assigning)
(input variable minimum value)
MaxX = 1;
(input variable maximum value)
MinX = 0;
(one of the possible membership function
parameters sets for each term of the output
variable)
A[10] = 0,3; B[14] = 0,8;
A[6] = 0,5; B[10] = 0,7;
A[18] = 0,4; B[22] = 0,95;
A[22] = 0,35; B[26] = 0,6;
(The number of points for numerical
integration)
IntCount = 100;
```

```
dx = (MaxX - MinX)/IntCount;
( To check the value of the input vector)
action = [1, 1, 1, 1];
end;
```

The next procedure is the use of the deductions from fuzzy logics rules. This procedure is called activation.

Let us use the method of prod-activation. In this method the resulting function is got by the means of multiplying the degree of the deduction validity by the corresponding membership function.

Let us make activation and accumulation simultaneously – find a common membership function for all terms. We make a common function for the set range of values. With this aim in purpose, the membership function of each term $\mu_i(X)$ should be multiplied by the value of the corresponding deduction validity degree (activation); it is also necessary to find the maximum value from all these products for each term (accumulation). Thus, we get an envelope function. The algorithm of calculation is given below.

(We make the accumulation of all terms into one function, which possesses maximum values out of each term membership functions).
```
function AccProb (X);
    AccProd (X) = action [1]* TriangleFM(X,
A[1], B[1]);
    for (i = 2,N)
        begin
        AccProb = max (AccProb,
action[i]*TriangleFM(X, A[1],B[1]);
        end;
end;
```

176

With the knowledge of the resulting membership function it is possible to find the obtained figure centre of mass.

On the basis of the input variables measured values, following the compositional deduction rules, we can determine possible permissible values of the air consumption. Fig. 6 presents graphical interpretation of this process for two rules.

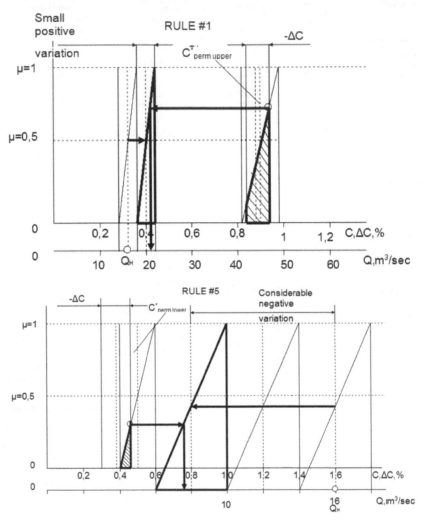

Figure 6. Deduction compositional rules graphical interpretation.

Fig. 7 presents the results of longwall goaf methane concentration control process by the means of air discharge variation simulation.

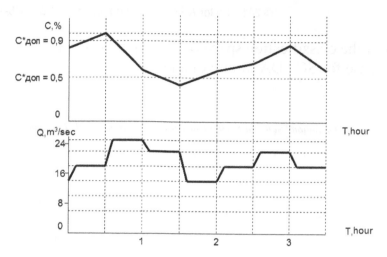

Figure 7. Driving signal and controlled variables variations graphics

3 CONCLUSIONS

The algorithm of this control is designed with the use of fuzzy logics. The rules of the controller functioning are based on the long-term experience of such control objects operation. The gained experimental knowledge made it possible to design a simple and reliable controller satisfying the necessary validity of control. The use of this algorithm provides the controller to secure reliable methane concentration under the decrease of longwall air average quantity and the diminution of the output process by 15÷30%.

REFERENCES

Abramov F.A., Tyan R.B. 1973. *Methods and algorithms for centralized control and management of ventilation shafts* (in Russian). Kyiv: Naukova dumka: 184.
Boyko B.A. & Petrechuk A.A. 1972. *Mathematical description of the mine excavation site as an object of control ventilation* (in Russian). Proceedings of the universities: Mining Journal, Issue 10: 87 – 92.
Kruglov V.V., Dli M.I. & Golunov P.Yu. 2000. *Fuzzy logic and artificial neural networks* (in Russian). Moscow: Fizmatlit: 225.

Power Engineering, Control and Information Technologies in Geotechnical Systems – Pivnyak, Beshta
& Alekseyev (eds)
© 2015 Taylor & Francis Group, London, ISBN 978-1-138-02804-3

Formation of the classification space of the expert system knowledge base for SCADA failure diagnostics.

O. Syrotkina

State Higher Educational Institution "National Mining University", Dnipropetrovs'k, Ukraine

ABSTRACT: SCADA fault diagnostics model based on Expert Diagnostic System is proposed in this article. The task of identifying dependencies between diagnostic codes and SCADA failure types is decided in this model. Object-oriented design technology is used to formalize the problem. The conditions of the search area diagnosis formation for the diagnostic model classification space are defined. Identification criteria of the object class using features and also in case of information insufficiency are analyzed.

1 INTRODUCTION

SCADA (Supervisory Control And Data Acquisition) systems are widely spread at present time to solve the problems of technological process automation in various fields of industrial production like power engineering, military engineering, transport, etc (Stouffer 2006). They are multi-level, hierarchical distributed hardware and software complexes (HSC) that work in real time. SCADA application areas, as responsible appointment systems, their complexity and versatility cause increased requirements for reliability and fault tolerance of such systems (Shestakov 2007). There is an actual task to perform automatic self-diagnostics of SCADA system functioning in real time. Using the methodology of Real Time Expert Systems (RTES) allows decide the tasks of SCADA diagnostics during its functioning. This methodology applies to explore complex multivariable and hard formalize models of systems. It allows perform intellectual analysis of input data large streams (Jackson 2001).

2 FORMULATION THE PROBLEM

SCADA functioning diagnostic model based on Expert System (ES) to explore the formation mechanisms of diagnosis search area using finding dependences between Diagnostic Codes and SCADA Fault Types is proposed to consider in this paper.

3 MATERIALS FOR RESEARCH

SCADA Diagnosis Subsystem (DS) based on Expert System (ES) deduces a conclusion about SCADA technical state and its functional components and processes founded on input data analysis from SCADA in real time. These data are System Diagnostic Codes (SDC).

According to (Jackson 2001) we have:

$$ES = <DB, KB, IM>, \qquad (1)$$

where ES – Expert System; DB – Database, that contains basic components (i.e. terms and facts of subject area, their properties and relations between them); KB – Knowledge Base, that contains structure variants building rules using Database basic components; IM – Inference Machine, that sets the order of applying the rules from Knowledge Base.

Some terms of subject area can be defined.

System Diagnostic Codes (SDC) are the set of software diagnostic code errors.

$$C = \{c_1, c_2, \ldots, c_{n(C)}\}, \qquad (2)$$

where $n(C) = |C|$ is cardinality of set C. It corresponds to the number of diagnostic codes defined in the system.

System Failure Types (SFT) can be presented as set F.

$$F = \{f_1, f_2, \ldots, f_{n(F)}\}, \qquad (3)$$

where $n(F) = |F|$ is cardinality of set F. It corresponds to the number of failure types.

Set G represents relationship between SDC and SFT as correspondence ratio of some SDC set to certain failure type.

$$G \subset C \times F = \{(c, f) \mid c \in C, f \in F\}. \qquad (4)$$

Relationship G (4) as a correspondence between Failure Types and System Diagnostic Codes is shown in Tab.1.

Table 1. The correspondence between SFT and SDC

C \ F	f_1	f_2	...	f_j	...	$f_{n(F)}$
c_1	$\alpha_{1,1}$	$\alpha_{1,2}$	...	$\alpha_{1,j}$	...	$\alpha_{1,n(F)}$
c_2	$\alpha_{2,1}$	$\alpha_{2,2}$	...	$\alpha_{2,j}$	...	$\alpha_{2,n(F)}$
...	...	...	...	...	...	...
c_i	$\alpha_{i,1}$	$\alpha_{i,2}$	...	$\alpha_{i,j}$	...	$\alpha_{i,n(F)}$
...	...	...	...	...	...	...
$c_{n(C)}$	$\alpha_{n(C),1}$	$\alpha_{n(C),2}$	...	$\alpha_{n(C),j}$	...	$\alpha_{n(C),n(F)}$

where $\alpha_{i,j} \in \{0,1,"-"\}$. On this basis we have following dependencies:

$\alpha_{i,j} = 0$. Diagnostic Code c_i has not relation to Failure Type f_j.

$\alpha_{i,j} = 1$. Subset of SDC includes Diagnostic Code c_i. This subset defines Failure Type f_j.

$\alpha_{i,j} = "-"$. Relation between Diagnostic Code c_i and Failure Type f_j is not defined.

We can consider relation G on the example of Euler and Venn Diagram from the point of view of correspondence with data taken from Tab. 1. This relation is shown in Fig. 1.

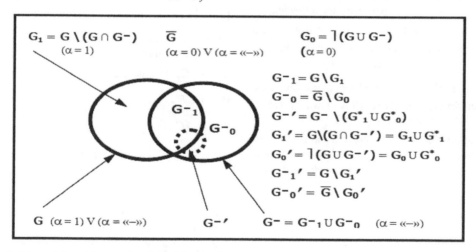

Figure 1. Euler and Venn Diagram for relation $G \subset C \times F$

To formalize the task for relation G between sets C and F we accept F value as a set of classes and C value as a set of features of these classes according to the object-oriented design (OOD) (Booch 2007).

According to this model:

G is partly defined area of features that belong to appropriate classes;

$\overline{G}$ is partly defined area of no-connection between features and classes.

So, for the current iterative step of Expert Diagnostic Real Time System we have following dependencies:

G_1 is area of appurtenance of features to classes (It is shown in Tab. 1. as $\alpha_{i,j} = 1$);

$$G_1 \subseteq G. \tag{5}$$

G_0 is area of no-connection between features and classes (It is shown in Tab. 1. as $\alpha_{i,j} = 0$).

$$G_0 \subseteq \overline{G}, \tag{6}$$

$$G_1 \cap G_0 = \varnothing. \tag{7}$$

$G-$ is uncertainty area. The appurtenance of features to classes is not defined (It is shown in Tab. 1. as $\alpha_{i,j} = "-"$).

$$G_1 = G \setminus (G \cap G-), \qquad (8)$$

$$G_0 = \overline{G} \setminus (\overline{G} \cap G-) = \overline{G \cup G-}. \qquad (9)$$

Uncertainty area $G-$ consists of the following elements:

$G-_1$ is area of potential appurtenance of features to classes;

$G-_0$ is area of potential no-connection between features and classes.

$$G- = G-_1 \cup G-_0, \qquad (10)$$

$$G-_1 = G \setminus G_1, \qquad (11)$$

$$G-_0 = \overline{G} \setminus G_0. \qquad (12)$$

As a result of Expert Diagnostic System work using $G-$ iterative step we have the following:

G_1^* is additionally defined area of features that belong to appropriate classes (It is shown in Tab. 1. as $\alpha_{i,j} = "-"$ changes on $\alpha_{i,j} = 1$);

G_0^* is additionally defined area of no-connection between features and classes (It is shown in Tab. 1. as $\alpha_{i,j} = "-"$ changes on $\alpha_{i,j} = 0$).

So, uncertainty area is decreased and for the next iteration step it becomes equal $G-'$ (It is shown in Tab. 1. as $\alpha_{i,j} = "-"$, i.e. $\alpha_{i,j}$ is not changed). As a result we have the following:

$$G-' = G - \setminus(G_1^* \cup G_0^*), \qquad (13)$$

$$G_1' = G \setminus (G \cap G-') = G_1 \cup G_1^*, \qquad (14)$$

$$G_0' = \overline{G \cup G-'} = G_0 \cup G_0^*, \qquad (15)$$

$$G-_1' = G \setminus G_1', \qquad (16)$$

$$G-_0' = \overline{G} \setminus G_0', \qquad (17)$$

where $G-'$ is uncertainty area for the next step of Expert Diagnostic System work; G_1' is area of appurtenance of features to classes for the next step of Expert Diagnostic System work; G_0' is area of no-connection between features and classes for the next step of Expert Diagnostic System work; $G-'$ is area of potential appurtenance of features to classes for the next step of Expert Diagnostic System work; $G-_0'$ is area of potential no-connection between features and classes for the next step of Expert Diagnostic System work.

There are following tasks of the proposed SCADA diagnostic model research:

- definition of conformity between SDC and SFT by detection the regularities in analyzing the implicative connections between features including prohibitions on the combinations of feature values (Zakrevsky 1988);

- definition of the minimum required set of features C_{min} to allow uniquely perform diagnose of Failure Type f.

Let us consider the task of formation the search area. We can imagine it as set-theoretic model to define connection between classification space and space of object states (Ustenko 2000).

Let the set of C_x^t with timestamp t that generates by using SCADA SDC is gone to the Expert Diagnostic System input.

$$C_x^t = \{c_1, c_2, \ldots, c_i, \ldots, c_n\}_x^t. \qquad (18)$$

Let Expert Diagnostic System Knowledge Base contains a set of rules $G_f \subset G$ that defines as a set of SDC C_f to make a diagnosis about having some SFT $f \in F$.

$$f = G_f(C_f), \qquad (19)$$

$$C_f = \{c_1, c_2, \ldots, c_i, \ldots c_n\}_f \subset C, \qquad (20)$$

$$G_f = \{g_1, g_2, \ldots, g_j, \ldots g_k\}_f \subset G. \qquad (21)$$

There is a following type of Skolem standard form of the deductive inference formula rules of Expert Diagnostic System Knowledge Base (Ponomarev 2005):

$$\vdash \overset{n}{\underset{i=1}{\forall}} c_i(\varphi(c_1, c_2, \ldots, c_n)) \leftarrow \overset{n}{\underset{i=1}{\exists}}(\overset{k}{\underset{j=1}{\wedge}} g_i(c_1, c_2, \ldots, c_n)). \qquad (22)$$

It is interpreted as: "We should prove the truth of conclusion hatchability from the set of true premises".

There are following designations in (22):

c_i is a variable that corresponds to some Diagnostic System Codes of Expert System Database;

g_j is a condition (premise) that corresponds to some rule of SFT definition;

φ is a conclusion (inference);

n is a number of variables;

k is a number of conditions.

Diagnosis search area can be defined as:

$$F_y = \{f_1, f_2, \ldots, f_n\}_y = \{$$

$$f \mid f = G_f(C_f), C_f \cap C_x^t \neq \varnothing\}. \qquad (23)$$

Let us consider possible variants of formation diagnosis search area F_y:

a) diagnosis search area F_a is fully defined on the input data set and contains only one solution;

$$\exists! f \subset G_f(C_f \subset C_x^t); \qquad (24)$$

b) diagnosis search area F_b is fully defined on the input data set and contains a set of alternative solutions (Solovyev 1996);

$$F_b \subset G_f(C_f \subset C_x^t); \qquad (25)$$

c) diagnosis search area F_c is partly defined on the input data set;

$$F_c \subset G_f(C_f), \ C_f \cap C_x^t \neq \varnothing; \qquad (26)$$

d) diagnosis search area F_d is not defined on the input data set;

$$F_d \subset G_f(C_f), \ C_f \cap C_x^t = \varnothing; \qquad (27)$$

e) diagnosis search area F_y consists of some combination we considered above in items a) ÷ d).

Accepting that set C_x has elements with equal significance coefficients to make a diagnosis, let's consider the system of rules for Expert Diagnostic System Knowledge Base. It is grouped by power set elements 2^{C_x} as shown in Example 1.

Example 1.

We assume the following:

C is a set of SDC that is in the Expert Diagnostic System Database;

G_1 is a system of Knowledge Base rules to detect SFT. It is formed for the current stage of Expert Diagnostic System;

C_x^t is a set of Expert Diagnostic System input data for the timestamp t.

We have the following:

$$C = \{c_1, c_2, c_3, c_4, c_5, c_6, c_7, c_8, c_9, c_{10}, c_{11}, c_{12}, c_{13}, c_{14}, c_{15}\},$$

$$G_1 = \{g_1, g_2, g_3 \cdot g_4, g_5, g_6, g_7, g_8, g_9, g_{10}\},$$

$$f_1 = g_1(c_1, c_4, c_6, c_{11}) = G_{f_1}(C_{f_1}),$$

$$f_2 = g_2(c_2, c_5, c_9, c_{11}, c_{15}) = G_{f_2}(C_{f_2}),$$

$$f_3 = g_3(c_2, c_3, c_7, c_{10}) = G_{f_3}(C_{f_3}),$$

$$f_4 = g_4(c_1, c_3, c_7) = G_{f_4}(C_{f_4}),$$

$$f_5 = g_5(c_7, c_{13}, c_{14}, c_{15}) = G_{f_5}(C_{f_5}),$$

$$f_6 = g_6(c_4, c_8, c_{12}, c_{13}) = G_{f_6}(C_{f_6}),$$

$$f_7 = g_7(c_5, c_8, c_{15}) = G_{f_7}(C_{f_7}),$$

$$f_8 = g_8(c_2, c_3, c_8, c_{14}) = G_{f_8}(C_{f_8}),$$

$$f_9 = g_9(c_6, c_{11}, c_{12}, c_{15}) = G_{f_9}(C_{f_9}),$$

$$f_{10} = g_{10}(c_9, c_{10}, c_{13}) = G_{f_{10}}(C_{f_{10}}),$$

$$C_x = \{c_1, c_3, c_7, c_{12}\}.$$

Find: diagnosis search area for the input data set C_x.

Let us collect our results of formation m-tuple Y_m^n (Gorbatov 2005) based on the input data set C_x in a Tab.2.

Table 2. m-tuples based on the input data set C_x

m	Y_m^n	$\lvert Y_m^n \rvert$
1	$Y_1^4 = \{c_1, c_3, c_7, c_{12}\}$	4
2	$Y_2^4 = \{(c_1, c_3), (c_1, c_7), (c_1, c_{12}), (c_3, c_7), (c_3, c_{12}), (c_7, c_{12})\}$	6
3	$Y_3^4 = \{(c_1, c_3, c_7), (c_1, c_3, c_{12}), (c_1, c_7, c_{12}), (c_3, c_7, c_{12})\}$	4
4	$Y_4^4 = \{(c_1, c_3, c_7, c_{12})\}$	1

For each element $y_{m,j}^n \in Y_m^n \subset 2^{C_x}$ formed on the Expert Diagnostic System input data set C_x correspondences the set of rules for Expert Diagnostic System Knowledge Base $G_{m,j}(y_{m,j}^n)$. It forms diagnosis search area $F_{m,j}$.

$$G_{m,j} : y_{m,j}^n \to F_{m,j},$$

$$F_{m,j} = \{f_1,\ldots,f_N\}_{m,j} = \{G_{f_1}(C_{f_1}),\ldots,G_{f_N}(C_{f_N})\}_{m,j},$$

$$C_{f_i} \cap 2^{C_x} = y_{m,j}^n,$$

$$C-_{f_i} = C_{f_i} \setminus y_{m,j}^n,$$

where $C-_{f_i}$ is required set of SDC to make a diagnosis f_i. Its element values is not defined on the input data set C_x.

If $C-_{f_i} = \varnothing$, then for the current state of Expert Diagnostic System Knowledge Base the set of rules G_{f_i} is fully defined on the input data set C_x, else it is partly defined.

Let us collect our results of forming alternative diagnosis search areas in a Tab.3.

Table 3. Alternative diagnosis search areas

No	$y_{m,j}^n$	F_y	$C-_{f_i}$
1	$y_{1,1}^4 = c_1$	f_1	$C-_{f_1} = \{c_4, c_6, c_{11}\}$
2	$y_{1,2}^4 = c_3$	f_8	$C-_{f_8} = \{c_2, c_8, c_{14}\}$
3	$y_{1,3}^4 = c_7$	f_5	$C-_{f_5} = \{c_{13}, c_{14}, c_{15}\}$
4	$y_{1,4}^4 = c_{12}$	f_6	$C-_{f_6} = \{c_4, c_8, c_{13}\}$
		f_9	$C-_{f_9} = \{c_6, c_{11}, c_{15}\}$
5	$y_{2,4}^4 = (c_3, c_7)$	f_3	$C-_{f_3} = \{c_2, c_{10}\}$
6	$y_{3,1}^4 = (c_1, c_3, c_7)$	f_4	$\varnothing$
7	$y_{2,1}^4 = (c_1, c_3)$		
	$y_{2,2}^4 = (c_1, c_7)$		
	$y_{2,3}^4 = (c_1, c_{12})$	$\varnothing$	$\varnothing$
	$y_{2,5}^4 = (c_3, c_{12})$		
	$y_{2,6}^4 = (c_7, c_{12})$		
	$y_{3,2}^4 = (c_1, c_3, c_{12})$		
	$y_{3,3}^4 = (c_1, c_7, c_{12})$		
	$y_{3,4}^4 = (c_3, c_7, c_{12})$		
	$y_{4,1}^4 = (c_1, c_3, c_7, c_{12})$		

So, there is following diagnosis search area for the current iterative step of Expert Diagnostic System (as it is shown in Tab. 3.):

$$F_y = F_a \cup F_c = \{f_4, f_6, f_9\}.$$

4 CONCLUSIONS

There are priority rules with fully defined parameters on the elements of power set 2^{C_x} with maximum power, that are required to make a diagnosis. Formation of the diagnosis search area by using the rules of Expert Diagnostic System Knowledge Base with partly defined parameters is performed by searching the addition from power set 2^{C_x} element to some subset of the ordered basic set of any power.

This approach in formation of diagnostic model classification space provides the efficiency of intellectual operations, and it reduces the time of performing of conflict resolution procedures for the set of alternative decisions. Additionally, it allows easy reconfigure knowledge formal structures that can significantly reduce the time to make a diagnosis about SCADA technical state.

REFERENCES

Stouffer K., Falco J., Kent K. 2006. *Guide to supervisory control and data acquisition (SCADA) and industrial control systems security. Recommendations of the National Institute of Standards and Technology.* NIST Special Publication 800-82: 164.

Shestakov D.A. 2007. *Requirements to modern software complexes control of integrated security systems* (in Russian). Security systems, Issue 1: 114-116.

Jackson P. 2001. *Introduction to Expert Systems* (in Russian). Moscow: Williams: 624.

Booch G., Maksimchuk R.A., Engle M.V., Young B.J., Conallen J., Houston K.A. 2007. *Object-oriented analysis and design with applications.* Boston: Addison-Wesley: 717.

Zakrevsky A.D. 1988. *Logic Recognition* (in Russian). Minsk: Science and technique: 118.

Ustenko A.S. 2000. *Principles of mathematical modeling and algorithmic processes of complex systems* (in Russian). Moscow: Binom: 250.

Ponomarev V.F. 2005. *Mathematical Logic* (in Russian). Kaliningrad: KSTU: 201.

Solovyev S.Yu. 1996. *Mathematical methods and principles of automated engineering systems building* (in Russian). PhD dissertation. [See http://park.glossary.ru/serios/read_05.php]

Gorbatov V.A. 2000. *Fundamentals of discrete mathematics. Information mathematics* (in Russian). Moscow: Science. Fizmatlit: 544.

© 2015 Taylor & Francis Group, London, ISBN 978-1-138-02804-3

Design of subordinate control system with two degrees of freedom

V. Sosedka, R. Mazur & M. Isakova
State Higher Educational Institution "National Mining University", Dnipropetrovs'k, Ukraine

ABSTRACT: The method of design of two-circuit system of subordinate control with two degrees of freedom is offered, one of which is used to design compensator of a main controlled variable, the other – to attribute preset disturbance-stimulated properties to a system. Discrete and continuous control system for induction motor variable-frequency vector induction electric drive of a vibrating feeder is designed according to the offered method. The performance of the system and its conformance to technical requirements is proved by simulation technique. The results of the research confirm the possibility of designing a system taking into account independent requirements on control and disturbance.

1 INTRODUCTION

In mining and metallurgy industry many technological processes involve vibrating machines and mechanisms such as vibrating screens, crushers, feeders, conveying chutes etc. Mass-production specialized induction motors with unbalanced disc are widely used as a source of mechanical vibration for such mechanisms. It is well-known that the highest efficiency of such vibrating mechanisms is achieved at near resonant working modes. This working mode ensures regulation and maintenance of optimal rotation speed of unbalanced discs depending on the load change. This problem is solved by using variable-frequency electric drives.

2 PROBLEM STATEMENT

Overwhelming majority of variable-frequency electric drives available on the market is based on unified systems designed on the principle of subordinate control. Serial connection of control regulators is a characteristic feature of this principle: the number of control regulators is equal to the number of controlled parameters of the electric drive (degree of freedom of control systems) such as current, magnetic flux, angular velocity, motion etc. Structural diagram of a typical dual loop system based on the principle of subordinate control is shown in Fig. 1.

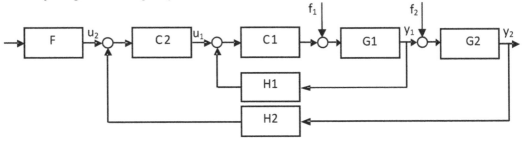

Figure 1. Structural diagram of a typical dual loop system

Such system includes compensators of internal C1 and external C2 loops, serial connection of plants represented as series connected elements G1 and G2 of the controlled system, sensors H1 and H2. Design of the system compensators of subordinated control system is done in a sequential order from internal circuit to the external one. Technical (modular) and symmetrical optimums are standard criteria of compensators synthesis in electric drives. In the majority of mass-production frequency converters the values of compensator parameters are set automatically in the course of automatic motor data identification procedure. Automatic motor data identification procedure consists of at least two stages. The first stage is realized with stalled rotor and implies determining the parameters of equivalent circuit of the electric drive with further calculation of current/flux controller parameters. The second stage is necessary to determine the parameters of speed controller and is realized with rotating engine. Realization of the second state in case of vibration mechanisms is usually impossible because of the likelihood of the construction failure, and thus prohibition of this operation under process specifics. In this case parameters of speed controller are set by setup staff during commissioning, which does not ensure required accuracy of speed control, and as a consequence, results in lower efficiency of vibration mechanisms.

3 RESEARCH TOOL

The research is done with imitation model of vector induction motor electric drive with modulated voltage-source inverter (Krause, Wasynczuk, Sudhoff 1995). Structural parameters of a vibrating channel REA 60/500 with unbalanced induction motor electric drive JVM JV206-850 (JÖST GmbH + Co. KG, Germany) installed at the site with continuous feed of bulk material into electric arc furnace by steel-making company JSC "MZ "Dneprosteel" are taken as input data for the research. The parameters of T-shaped equivalent circuit diagram necessary for the calculation are obtained through motor data identification by frequency converter Sinamics G120 with processor unit CU240E-2DP 6SL3244-0BB12-1PA1 and power module PM240-2 6SL3210-1PE18-0AL1 (SIEMENS AG, Germany).

The structural diagram of the model is presented in Fig.2.

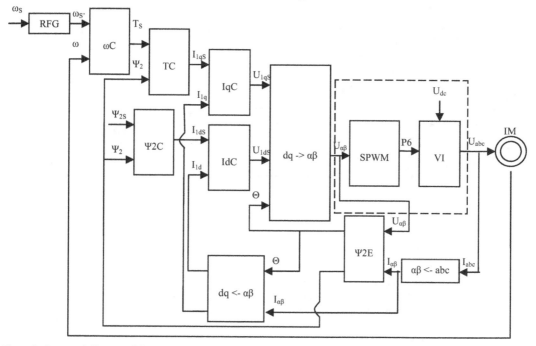

Figure 2. Structural diagram of the model

In the structural diagram (Fig. 2): 'IM' is the model of induction machine with short circuit squirrel cage rotor; 'VI' is a bridge with 6-pulse modulated voltage-source inverter powered by a DC power supply Udc; P6 are driving pulses generated by space pulse-width modulation SPWM (Trzynadlowski, Kirlin, Legowski 1997); "abc->αβ", "αβ -> dq", "dq->αβ" are the Park and Clarke coordinates transformers; Ψ2E is the estimator of generalized vector of rotor flux; IdC и IqC are controllers of flux and torque components of stator current; Ψ2C is the rotor flux controller; TC is the torque controller; ωC is the velocity controller. Setpoints of controlled variables are indicated with s index.

The model of induction machine IM is based on the well-known expressions for generalized electric machine in the rotating coordinate system (Pivnyak, Volkov 2006):

$$
\begin{cases}
U_{1d} = R_1 I_{1d} + \left(L_1 - \dfrac{L_0^2}{L_2}\right)\dfrac{dI_{1d}}{dt} + \dfrac{L_0}{L_2}\dfrac{d\Psi_{2d}}{dt} - \omega_{cs}\left(L_1 - \dfrac{L_0^2}{L_2}\right)I_{1q}; \\[3mm]
U_{1q} = R_1 I_{1q} + \left(L_1 - \dfrac{L_0^2}{L_2}\right)\dfrac{dI_{1q}}{dt} + \omega_{cs}\left(L_1 - \dfrac{L_0^2}{L_2}\right)I_{1q} + \omega_{cs}\dfrac{L_0}{L_2}\Psi_{2d}; \\[3mm]
0 = \dfrac{R_2}{L_2}\Psi_{2d} - L_0\dfrac{R_2}{L_2}I_{1d} + \dfrac{d\Psi_{2d}}{dt}; \\[3mm]
0 = -L_0\dfrac{R_2}{L_2}I_{1q} + (\omega_{cs} - p_p\omega)\Psi_{2d}; \\[3mm]
T = \dfrac{3}{2}\cdot\dfrac{p_p L_0}{L_2}\cdot I_{1q}\cdot\Psi_{2d},
\end{cases}
$$

where U_{1d}, U_{1q} are projections of generalized vector of stator voltage plotted on the orthogonal axis rotating with the velocity ω_{cs} of the system of axes; I_{1d}, I_{1q} are the projections of the generalized vector of the stator current; Ψ_{2d}, Ψ_{2q} – projections of generalized vector of rotor flux linkage; R_1, R_2 – active resistances of stator and rotor winding of the given generalized machine; L_1, L_2 – complete self-inductances of stator and rotor windings; L_0 – mutual inductance; p_p – the number of pole pairs; ω – rotor angular spin rate; T – electromagnetic torque. These equations correspond to the detailed structural diagram of an asynchronous machine as a control object given in Fig. 3.

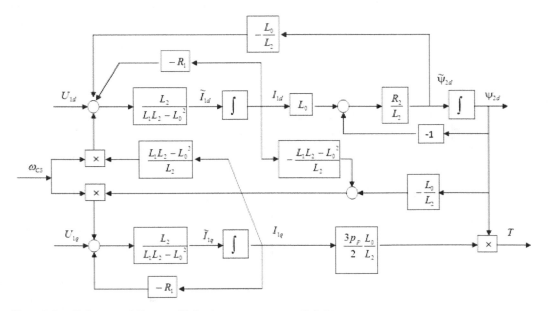

Figure 3. Detailed structural diagram of induction motor as a controlled object

We design a control device for the given object in accordance with the principles of subordinate control. Control loops for flux I_{dx} and torque I_{qy} components of stator current are internal in the system. Control loop of flux-component of starting motor current I_{1d} is shown in Fig. 4. Once the loop is set for modular optimum, "IdC" current controller transfer function will have the form:

$$
W_{IdC}(s) = \frac{R_1}{2T_o K_{VI} K_i}\left(T_{12} + \frac{1}{s}\right),
$$

where "s" is Laplace domain variable; "K_{VI}" is a gain of space pulse width modulated voltage inverter; Ki is the current feedback gain; "T_{12}" is the transient time constant of the stator circuit:

$$
T_{12} = \frac{L_{12}}{R_1} = \frac{L_1 L_2 - L_0^2}{L_2 R_1};
$$

"T_o" is an uncompensated time constant:

$$
T_o \approx 0.2 T_{12}.
$$

Structural diagram of the control loop of torque component of stator current I1q is similar to one given in Fig. 4, the parameters of the "IqC" control is calculated similarly.

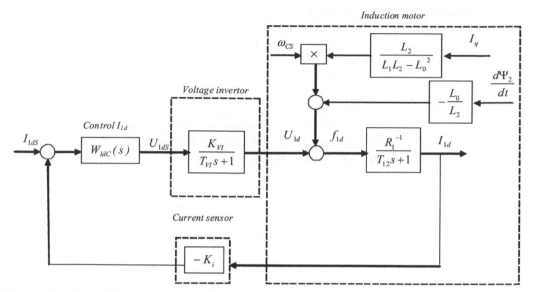

Figure 4. Control loop of flux component of stator current I_{1x}

Transfer function of a closed current loop set for modular optimum has the standard form of second order filter $W_I(s) = \dfrac{K_i^{-1}}{2T_0^2 s^2 + 2T_0 s + 1}$, which with general assumption $2T_0^2 s^2 \rightarrow 0$ taken into account

can be reduced to the first order filter with transfer function $W_I(s) \approx \dfrac{K_i^{-1}}{2T_0 s + 1}$.

Structural diagram of an external control loop of flux component of a rotor Ψ_{2x} with the above considerations taken into account is presented in Fig. 5.

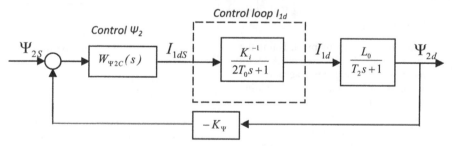

Figure 5. Control loop of flux component of rotor $\Psi 2x$

Based on the astatism requirement to the control loop of rotor flux component, the grounded criterion of control design is the setting of the loop for symmetrical optimum, from whence the transfer function of flux component control is:

$$W_{\psi 2C}(s) = \frac{K_i\left(T_2 + \dfrac{1}{s}\right)}{4T_0 L_0 K_\psi},$$

where $T_2 = \dfrac{L_2}{R_2}$ is the electro-magnetic time constant of the rotor circuit; K_Ψ is the controlled parameter feedback gain.

It should be noted that in the model (Fig. 2) rotor flux linkage is not measured directly and is calculated according to formulation (block "$\Psi 2E$"):

$$\overline{\Psi}_2 = |\Psi_2| e^{j\Theta} = \sqrt{\Psi_{2\alpha}^2 + \Psi_{2\beta}^2}\, e^{j arctg \frac{\Psi_{2\beta}}{\Psi_{2\alpha}}},$$

188

where projections of generalized vector of rotor flux linkage on orthogonal axis of fixed coordinates are:

$$\psi_{2\alpha,\beta}(s)=\frac{L_2}{L_0}\frac{-(L_1-\frac{L_0^2}{L_2})I_{1\alpha,\beta}s+U_{1\alpha,\beta}-I_{1\alpha,\beta}R_1}{s}.$$

The last expression is easy to obtain from the equations of generalized asynchronous machine.

Connection between three-phase (abc), two-phase fixed (αβ) and rotating (dq) coordinates is determined according to Park and Clarke formulas (blocks "abc->αβ", "αβ -> dq", "dq->αβ"):

$$\begin{cases}V_\alpha=V_A;\\ V_\beta=\frac{1}{\sqrt{3}}V_B-\frac{1}{\sqrt{3}}V_C,\end{cases}\quad\begin{cases}V_d=V_\alpha\cos\Theta+V_\beta\sin\Theta;\\ V_q=-V_\alpha\sin\Theta+V_\beta\cos\Theta,\end{cases}$$

$$\begin{cases}V_\alpha=V_d\cos\Theta-V_q\sin\Theta;\\ V_\beta=V_d\sin\Theta+V_q\sin\Theta.\end{cases}$$

Torque control group (block "TC") formulates the setpoint on the torque component of stator current according to algebraic expression:

$$I_{1qS}=T_S\frac{2L_2}{3p_pL_0}\frac{1}{\Psi_2}.$$

Structural diagram of velocity control loop with optimized control loop of torque component of current stator is presented in Fig. 6.

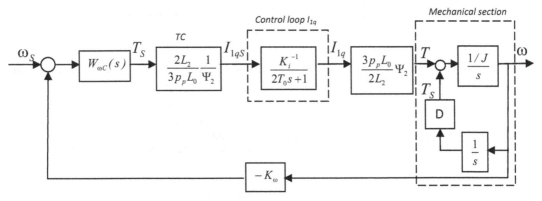

Figure 6. Velocity control loop

In Fig.6: J is the motor inertia moment; Kω is the velocity feedback gain; D is the block with nonlinear transfer gain taking into account the change of static torque on the motor shaft depending on the position of unbalanced disc; $W_{\omega C}(S)$ is the transfer function of velocity control.

Setting T_s by disturbing action, from the conditions of velocity loop settings for modular optimum, we obtain proportional control with transfer function $W_{C\omega}{}^{MO}(s)=\dfrac{K_iJ}{4T_0K_\omega}$. In practice, as a rule, gain of proportional velocity controller is set experimentally during vibration mechanisms commissioning.

4 RESULTS OF INVESTIGATION

It should be noted that as the transfer function of mechanical section of unbalanced induction motor

$W_M(s)=\dfrac{\omega(s)}{T(s)}=\dfrac{s}{Js^2+D}$ is nonlinear (Fig.6), the

quality of velocity control with a traditionally designed control does not always meet process requirements.

There arise the issues of improving the system characteristics and the price and method of their improvement.

One of the main characteristics of closed loop of closed automation system is its ability to reduce its sensitivity to variations of the system parameters. To quantitatively evaluate the variations, the sensitivity function S(s) is introduced to connect infinitely small relative variations of transfer function of the controlled object with relatively infinitely small variations of the transfer function of the closed system. As an example, we mathematically define the function of sensitivity to parameters of the control object with respect to internal loop of the system presented in Fig. 1:

189

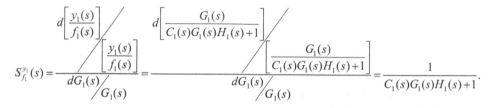

$$S_{u_1}^{y_1}(s) = \frac{d\left[\dfrac{y_1(s)}{u_1(s)}\right] \Big/ \dfrac{y_1(s)}{u_1(s)}}{dG_1(s) \Big/ G_1(s)} = \frac{d\left[\dfrac{C_1(s)G_1(s)}{C_1(s)G_1(s)H_1(s)+1}\right] \Big/ \dfrac{C_1(s)G_1(s)}{C_1(s)G_1(s)H_1(s)+1}}{dG_1(s) \Big/ G_1(s)} = \frac{1}{C_1(s)G_1(s)H_1(s)+1}.$$

The sensitivity function to disturbance of the object is calculated similarly:

$$S_{f_1}^{y_1}(s) = \frac{d\left[\dfrac{y_1(s)}{f_1(s)}\right] \Big/ \dfrac{y_1(s)}{f_1(s)}}{dG_1(s) \Big/ G_1(s)} = \frac{d\left[\dfrac{G_1(s)}{C_1(s)G_1(s)H_1(s)+1}\right] \Big/ \dfrac{G_1(s)}{C_1(s)G_1(s)H_1(s)+1}}{dG_1(s) \Big/ G_1(s)} = \frac{1}{C_1(s)G_1(s)H_1(s)+1}.$$

As follows from the analysis of the obtained expression, at $C_1(s)G_1(s)H_1(s) \gg 1$ the sensitivity functions to the object parameters

$$S_{u_1}^{y_1}(s) = S_{f_1}^{y_1}(s) \approx \frac{1}{C_1(s)G_1(s)H_1(s)}$$

are inversely proportional to the closed control loop gain.

Assuming that in general case the system sensitivity is $M_s = \lim_{0 \le \omega < \infty} |S(j\omega)|$, it is convenient to evaluate it by magnitude of Bode diagram. Let us consider Bode diagram of the control loops of torque and flux component of stator current that are set for module optimum. According to the criteria of setting an open loop for module optimum

$$W_{MO}(s) = \frac{1}{2T_o s(T_o s + 1)}$$ Bode diagram will take the

form presented in Fig. 7 – curve 1.

As sensitivity function is a kind of frequency characteristic, it is easy to evaluate by Bode diagram. From Bode diagram of open-loop system (Fig.7) it follows that sensitivity at low frequencies is also low. As the frequency grows sensitivity increases, and in the frequency range over crossover frequency $1/(2T_o)$ the system becomes very sensitive to the plant parameter variation. Feedback coupling in this frequency range becomes "positive" as the sensitivity of forward loop increases in comparison with these blocks without feedbacks. In case when the system is synthesized for module optimum, damping coefficient is equal to $\sqrt{2}/2 \approx 0.707$. In such system, frequency range with "positive" feedback makes 30% of bandpass. It should be noted that the above given definitions of sensitivity are accurate enough for practical use if the variations of block parameters of forward loop do not exceed 30%. In case when block parameters have greater variations, the definition of sensitivity by Horowitz should be applied.

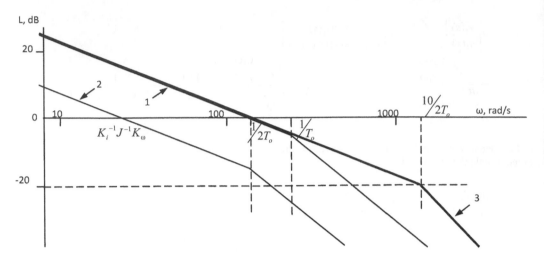

Figure 7. Bode diagram (1 is the current loop set for module optimum; 2 is the open velocity loop without correcting device; 3 is a desired characteristic of velocity loop).

Let us consider Bode diagram synthesis for module optimum of control loop of torque and flux components of stator current. According to criterion of synthesis for module optimum

$$W_{MO}(s) = \frac{1}{2T_o s(T_o s + 1)}$$ open loop Bode diagram

takes the form presented in Fig.7 – curve 1. Analysis of Bode diagram (Fig.7) shows that to reduce sensibility of loop set for module optimum it is necessary to increase amplification factor in the frequency range $\frac{1}{2T_o} \leq \omega \leq \frac{1}{T_o}$ at simultaneous adjustment of phase shift (without reducing phase margin). This transformation of Bode diagram demands for considering stability not with regard to boundary point but to boundary region. In classical control theory it is commonly supposed that boundary-point stability allowing to design systems with standard controllers with proportional, proportional-integral and proportional-integral-differential actions. For the system to have higher quality characteristics than at standard settings, the system in the range of essential frequencies has to have greater value of amplification factor, and the control has to have a more complicated transfer function that would ensure passing of magnitude-phase characteristic of optimized system along the boundary of stability range (Lurie, Enright 2000).

Let us apply these considerations to unbalanced asynchronous drive and consider structural diagram, Figs. 4-6. Despite the fact that asynchronous drive in general case should work out the controlling and disturbing influences, the control is designed in accordance with control requirements, and

disturbance transfer function is obtained automatically as a result of control system optimization. That is, the system has two inputs and potentially two degrees of freedom, but to ensure the preset properties there is used only one degree of freedom with respect to control. In case of only one degree of freedom, it is impossible to obtain performance limits with respect to both control and disturbance simultaneously.

The method of frequency (logarithmic) characteristics theoretically allows an engineer to find a compromise with respect to both problems. The technique of two-circuit system of control with two degrees of freedom is reduced to plotting of Bode diagram of closed internal loop set for module optimum. Then there is constructed the Bode diagram of open external loop set for module or symmetrical optimum. The obtained characteristic is corrected in the medium frequency range to reduce the sensitivity and ensure the required quality of disturbance transient. After that on the basis of Bode diagram the transfer function of series correcting device is calculated with traditional method.

We illustrate the offered method with the following example of synthesis a system of automatic control of vector variable-frequency electric drive with unbalanced asynchronous electric motor. Let the internal loop of torque component of stator current is set for module optimum and has Bode diagram given in Fig. 7 – curve 1. Then taking into account Fig.6, transfer function of Bode diagram of open speed loop will take the form of

$$W_{GH\omega}(s) = \frac{K_i^{-1} J^{-1} K_\omega}{s(2T_0 s + 1)}.$$ This transfer function is

correspondent to magnitude Bode diagram given in Fig. 7 – curve 2. Let us formulate the requirements to the speed loop by disturbance sensitivity. Let's assume that in the medium frequency range the magnitude of Bode diagram is not allowed lower than -20 dB. As the medium frequency range ensures the quality of transient process, then for approaching to its nonperiodic view, this range incline should be -20 dB/dec, and its length should be a decade instead of an octave. Let us assume that in the high frequency range, the magnitude of Bode diagram inclination is -60 dB/decade. Curve 3 (Fig.7) meets these requirements. According to the type of the desired magnitude of Bode diagram we obtain the desired transfer function of open speed loop $W_{CGH\omega}(s) = \dfrac{1}{2T_0 s (0.2T_0 s + 1)^2}$, from which transfer function of series correcting device is:

$$W_{C\omega}(s) = \frac{K_i J}{2T_0 K_\omega} \frac{2T_0 s + 1}{(0.2T_0 s + 1)^2} \approx \frac{K_i J}{2T_0 K_\omega} \frac{2T_0 s + 1}{0.4T_0 s + 1}.$$

Comparative analysis of transfer functions of the offered and traditional proportional controls shows that to reduce the system disturbance sensitivity it is enough to connect in series the element with transfer function $W_{C2\omega}(s) = 2\dfrac{2T_0 s + 1}{0.4T_0 s + 1}$ along with traditional proportional controller.

By substituting in the last expression $s = \dfrac{2(z-1)}{T_S(z+1)}$, we obtain z-transfer function with series correcting device

$$W_{C2\omega d}(z) = 2\frac{(4T_0 + T_S)z - 4T_0 + T_S}{(0.8T_0 + T_S)z - 0.8T_0 + T_S},$$ where T_S is discrete system sample time.

The results of modeling systems designed by traditional and offered methods are presented in graphs in Fig.8. Unbalanced motor JVM JV206-850 was taken as the controlled object with the following characteristics of T-shaped equivalent circuit: stator active resistance is R_1=1.8471 Ohm; rotor active resistance is R_2=1.74456 Ohm; stator leakage inductance L_{1l}=20.5 mH; rotor leakage inductance L_{2l}=23 mH; mutual inductance L_0=193.3 mH. The amplification factors gains of modulated by SPWM voltage-source inverter K_{VI}, feedbacks K_i, K_Ψ, K_ω are assumed to equal a unit. The transient time constant of stator circuit is

$$T_{12} = \frac{L_{12}}{R_1} = \frac{L_1 L_2 - L_0^2}{L_2 R_1} = \frac{(L_{1l} + L_0)(L_{2l} + L_0) - L_0^2}{(L_{2l} + L_0)R_1} = 22\,\text{ms/rad}^,$$

uncompensated fast time constant is T_0=4 ms/rad. Transfer function of current controllers is

$$W_{Id,qC}(s) = \frac{R_1}{2T_0 K_{VI} K_i}\left(T_{12} + \frac{1}{s}\right) = 5 + \frac{230}{s},$$ and of flux linkage is $W_{\psi 2C}(s) = \dfrac{K_i\left(T_2 + \dfrac{1}{s}\right)}{4T_0 L_0 K_\Psi} = 40 + \dfrac{320}{s}.$

Proportional speed controller gain set during commissioning the vibrating channel is $W_{C\omega}{}^{MO}(s) = 200$. Continuous and discrete transfer functions of the offered series correcting device of speed loop are

$$W_{C2\omega}(s) = 2\frac{2T_0 s + 1}{0.4T_0 s + 1} = 2\frac{0.008s + 1}{0.0016s + 1},$$

$$W_{C2\omega d}(z)\Big|_{T_S = 0.25 \cdot 10^{-3}c} = 2\frac{(4T_0 + T_S)z - 4T_0 + T_S}{(0.8T_0 + T_S)z - 0.8T_0 + T_S} =$$
$$= 2\frac{0.0163z - 0.0158}{0.0034z - 0.003}.$$

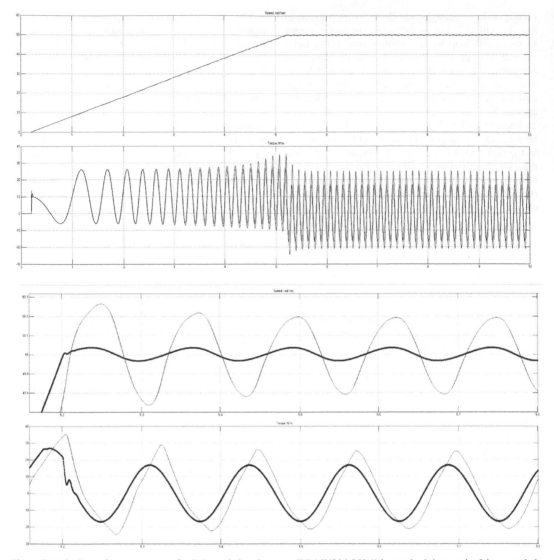

Figure 8. Velocity and torque curves of unbalanced electric motor JVM JV206-850 (1 is standard time scale; 2 is expanded time scale.

The analysis of curves in Fig.7 shows that introduction of the offered corrective device will allow reducing velocity control error by a factor of 4, and at the same time reduce the vibration amplitude of the electric drive torque. Thus, taking into account the second degree of freedom when designing automated control systems allows reducing control error, while the parameters of the additional correcting device introduced into the control direct circuit are calculated based on the equivalent circuit data of the electric motor, moreover they do not require testing on moving vibration mechanisms.

The authors consider that the obtained correcting device in the given example is only one of many possible solutions that might be obtained through transformation of desired Bode diagrams.

CONCLUSIONS

1. Traditional design methods for velocity controls of variable-frequency electric drives with unbalanced asynchronous electric motors do not allow achieving maximal efficiency of vibration mechanisms because of the disturbing action nonlinearly connected with velocity and position of the drive shaft.

193

2. The reduction of control error is possible at the design of control system with two degrees of freedom taken into account. The offered method using two degrees of freedom allows determining the parameters of controls analytically or in diagram form with the use of Bode diagrams. The offered design method might be used in electric drives of vibration mechanisms, as it is based on parameters of T-shaped equivalent circuit of asynchronous electric motor it does not need testing on rotating motor; it also allows reducing disturbance sensitivity. According to the results of variable-frequency drive simulation, the series correcting device with standard proportional control allows reducing control error by the factor of 4 and more, which proves the expediency of using systems with two degrees of freedom.

3. The offered correcting device does not require the use of specialized computing devices and can be easily realized by microprocessor technology means.

REFERENCES

Krause P.C., Wasynczuk O., Sudhoff S.D. 1995. *Analysis of Electric Machinery.* IEEE Press: 564.

Trzynadlowski A.M., Kirlin R.L., Legowski S.F. 1997. *Space vector PWM technique with minimum switching losses and a variable pulse rate.* IEEE Transactions on Industrial Electronics, Vol. 44, Issue 2: 173-181.

Pivnyak G.G., Volkov A.V. 2006. *Modern variable speed asynchronous electric drive with PWM* (in Russian). Dnipropetrovs'k: NMU: 470.

Horowitz I. M. 1963. *Synthesis of feedback systems.* New York: Academic Press: 726.

Lurie B., Enright P. 2000. *Classical Feedback Control: With MATLAB.* CRC Press: 456.

Power Engineering, Control and Information Technologies in Geotechnical Systems – Pivnyak, Beshta
& Alekseyev (eds)
© 2015 Taylor & Francis Group, London, ISBN 978-1-138-02804-3

Anaerobic digestion model of solid wastes in small biogas plants: influence of natural convection on biogas production.

A. Zemlyanka
Biogas Research Centre, Dnipropetrovs'k, Ukraine

M. Gubynskyi
National Metallurgical Academy, Dnipropetrovs'k, Ukraine

T. Vvedenska
State Higher Educational Institution "National Mining University", Dnipropetrovs'k, Ukraine

ABSTRACT: The article presents the study of natural convection impact on biogas production. Hydrodynamic and heat transfer processes were studied as a part of anaerobic digestion process by way of numerical experiments. A possible error in biogas yield calculations was estimated on the basis of a more simplified approach compared with conventional mathematical modeling. Potential increase in biogas production due to natural convection was also the object of investigation.

1 INTRODUCTION

It is known (Khanal 2008; Zemlianka, Gubynskyy 2008) that one of the most important conditions for high biogas plant productivity is uniform distribution of temperature and reactant concentrations in the reactor space. Recent numerical experiments proved that temperature distribution is more important and can be achieved by recurred heating and efficient mixing (Zemlianka, Gubynskyy 2008). In order to conduct a more detailed analysis of a biomass heating process with the view to designing and optimizing biogas reactors mode numerical experiments were conducted, which allowed to estimate a possible error in biogas yield calculations assess potential biogas production increase for small biogas plants by controlling heat and mass transfer processes.

2 MATERIALS AND METHODS

A mathematical model described in (Zemlianka 2007) is used to analyze the influence of heat and mass transfer processes in a biogas digester. The equation studies heat and mass transfer in a cylindrical reactor in terms of natural convection (Fig. 1). The boundary conditions are set by constant heat flux. In addition, the model is developed by considering bioreactor operational mode i.e. heating of biomass within the temperature range 29-31°C for 20-30 minutes and

stirring it for 20 minutes three times a day every six hours. This model takes into account the influence of natural convection in addition to different heater location options. In order to decrease the number of experiments the heater location options on the reactor height are considered.

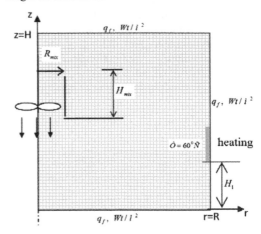

Figure. 1. Biogas reactor and its grid layout.

The following system of four equations to describes heat and mass transfer processes (Zemlianka 2004) in the 2-D biogas reactor model with cylindrical symmetry:

$$\frac{\partial u}{\partial \tau} + \frac{\partial (uu)}{\partial r} + \frac{\partial (vu)}{\partial z} + \frac{uu}{r} = v\left(\frac{\partial^2 u}{\partial r^2} + \frac{\partial^2 u}{\partial z^2} + \frac{1}{r}\frac{\partial u}{\partial r}\right) - \frac{1}{\rho}\frac{\partial p}{\partial r} \tag{1}$$

$$\frac{\partial v}{\partial \tau} + \frac{\partial (uv)}{\partial r} + \frac{\partial (vv)}{\partial z} + \frac{uv}{r} = v\left(\frac{\partial^2 v}{\partial r^2} + \frac{\partial^2 v}{\partial z^2} + \frac{1}{r}\frac{\partial v}{\partial r}\right) - \frac{1}{\rho}\frac{\partial p}{\partial z} + g\beta(\dot{O} - \dot{O}_i) + q_{mix} \tag{2}$$

$$\frac{\partial (u)}{\partial r} + \frac{\partial (v)}{\partial z} + \frac{u}{r} = 0 \tag{3}$$

$$\frac{\partial T}{\partial \tau} + \frac{\partial (uT)}{\partial r} + \frac{\partial (vT)}{\partial z} + \frac{uT}{r} = a\left(\frac{\partial^2 T}{\partial r^2} + \frac{\partial^2 T}{\partial z^2} + \frac{1}{r}\frac{\partial T}{\partial r}\right) \tag{4}$$

where r, z – radius and vertical coordinate of the cylinder, m; u, v – radial and vertical velocity components, m/sec; v – kinematic viscosity, m^2/sec; $g\beta(\dot{O} - \dot{O}_h)$ - the rate of liquid inner moment due to natural convection; g - free fall acceleration, m/sec^2; β - thermal expansion coefficient, 1/K; ρ - density, kg/m^3; T_h - temperature of the heater surface, °C; q_{mix} - the rate of movement generation (due to the stirring device) in each control volume, m/sec^2. a- thermal conductivity, m^2/sec.

The given heat and mass transfer model is to be completed together with a kinetic model which consists of four differential equations for the law of conservation of a chemical component. This kinetic model describes: hydrolysis of solid waste (5), substrate genesis (6), methanogenic biomass growth (7), methanogenesis (8):

$$\frac{\partial W}{\partial \tau} + \frac{\partial (uW)}{\partial r} + \frac{\partial (vW)}{\partial z} + \frac{uW}{r} = -kWf_h(S) \tag{5}$$

$$\frac{\partial S}{\partial \tau} + \frac{\partial (uS)}{\partial r} + \frac{\partial (vS)}{\partial z} + \frac{uS}{r} = D_S\left(\frac{\partial^2 S}{\partial r^2} + \frac{\partial^2 S}{\partial z^2} + \frac{1}{r}\frac{\partial S}{\partial r}\right) + \chi kWf_h(S) - \rho_m T_f f_m(S)\frac{SB}{K_S + S} \tag{6}$$

$$\frac{\partial B}{\partial \tau} + \frac{\partial (uB)}{\partial r} + \frac{\partial (vB)}{\partial z} + \frac{uB}{r} = D_B\left(\frac{\partial^2 B}{\partial r^2} + \frac{\partial^2 B}{\partial z^2} + \frac{1}{r}\frac{\partial B}{\partial r}\right) + Y_{X/S}\rho_m T_f f_m(S)\frac{SB}{K_S + S} - k_d B \tag{7}$$

$$\frac{\partial P}{\partial \tau} + \frac{\partial (uP)}{\partial r} + \frac{\partial (vP)}{\partial z} + \frac{uP}{r} = A(1 - Y_{X/S})\rho_m T_f f(\tau)f_m(S)\frac{SB}{K_S + S} \tag{8}$$

where W – solid waste concentration, g/l; S – total acid concentration, g/l; $\hat{A}$ – methanogenic biomass concentration, g/l; D – biogas concentration, g/l; T_f – dimentionless temperature function for methanogenic reactions. This kinetic model has been described in detail recently (Zemlianka 2007), so there's no need to consider it here. According to (Leontyev 1979), single-valuedness conditions for all equations of the system (1-4) consist in: boundary conditions – table 1; initial conditions – table 2; condition characterizing periodicity of stirring mode and heating mode - table 3.

Table 1. Boundary conditions for cylindrical bioreactor

Boundary	value	conditions
bottom of the reactor	$0 < r < R$, $z = 0$	$u, v = 0$; $q = q_l$;
free surface	$0 < r < R$, $z = \hat{I}$	$\dfrac{\partial u}{\partial z} = 0$; $v = 0$; $q = q_l$;
symmetry axis	$0 < z < H$, $r = 0$	$u = 0$; $\dfrac{\partial v}{\partial r} = 0$; $\dfrac{\partial T}{\partial r} = 0$;

Continuation of Table 1.

Boundary	value	conditions
interior side wall	$0 < z < H$, $r = R$	$u, v = 0$; $q = q_l$;
exterior surface of the heater	$0.2H < z < 0.35H$, $r = R - \Delta r$	$u, v = 0$; $T = T_h$.
stirring zone	$H_{mix} < z < H$, $0 < r < R_{mix}$	$\dfrac{\partial v}{\partial \tau} = q_{mix} = const$, $m^2 / \sec$; $\tau = \tau_n$

where q_l – heat loss per unit of the surface, Wt / m^2 ; λ – conductivity coefficient for liquid waste, $Wt /(m \cdot K)$.

Table 2. Initial conditions

equation	conditions
equations of movement	$u(r, z, 0) = 0$; $v(r, z, 0) = 0$;
equations of heat transfer	$\dot{O}(r, z, 0) = \dot{O}_0$;

Table 3. Conditions, governing the operation mode

Process	Conditions
loading	$\Delta \tau_{\varsigma} = 1$ time a day
unloading	same as loading
stirring	$\Delta \tau_{\ddot{i}} = 3$ times a day
heating	within temperature range 29-31°C

This mathematical simulation was carried out for biogas plant with 6 m^3 fermenter volume, fermenter radius $R = 1$ m ; fermenter height $H = 2$ m. A stirring device covered zone with $R_{mix} = 0.5R$ radius and $H_{mix} = 0.6H$ depth. Heat-transfer fluid temperature was 60°C, initial biomass temperature (T_0) constituted 29°C, for heat loss $q_l = 29$ Wt / m^2. Biomass thermophysical properties: kinematic viscosity $v = 0.000195$ $m^2 / \sec$; thermal conductivity $\dot{a} = 125.12 \cdot 10^{-5}$ $m^2 / \sec$.

Numerical experiments were conducted using the "staggered grid" with 20×20 points, the control volume numeric method and the algorithm SIMPLE (Kochubey, Ryadno 1991). This anaerobic digestion model was validated by kinetic part and heat transfer part separately (Kochubey, Ryadno 1991; Zemlianka 2004; 2007).

3 RESULTS AND DISCUSSION

The temperature field and the flow field for two-dimensional situation during heater operation are shown in Fig. 2. The heater was located according to the scheme shown in Fig.1 at a height of $z = 0.4H_1$ from the bottom of the reactor. According to the bioreactor operation mode, the data was registered at $\tau = 20$ min, which corresponds to the heater operation time. As can be seen from the temperature field and the flow field, the vertical fluid flow is generated next to the exterior wall of the heater due to natural convection. This vertical fluid flow has maximum velocity $v_{max} = 0.16$ $sm / \sec$ and produces some extra stirring of biomass.

Fig. 3 presents the temperature field and the flow field after the heater is turned off at $\tau = 2$ h, which corresponds to the period of time between heating and stirring of biomass. Analyzing the temperature field and the flow field shown in Fig.3, we can see that natural convection exerts irregular temperature field, with a zone of high temperature at the top of the reactor. At the same time, a cooling downstream is generated next to the interior wall of the reactor with maximum velocity $v_{max} = 0.014$ $sm / \sec$ which doesn't produce any significant effect on stirring of biomass.

Fig. 4 shows the results of a similar study. The

temperature field and the flow field were registered at $\tau = 6$ h which is the time for stirring device to turn on. As can be seen from the data shown in Fig. 4, an almost regular temperature field forms up during 20 min stirring which ensures high productivity of the bioreactor, near to maximum.

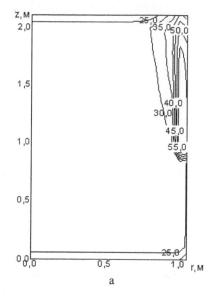

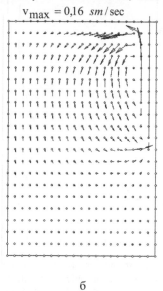

Figure 2. The temperature field and the flow field two-dimensional situation during heater operation at $\tau = 20$ min.

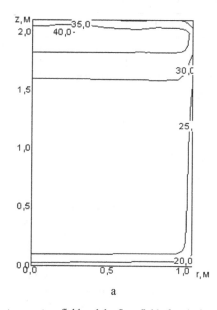

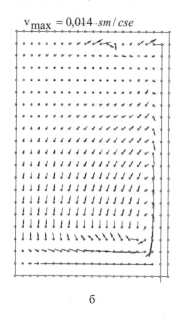

Figure 3. The temperature field and the flow field after the heater is turned off at $\tau = 2$ h.

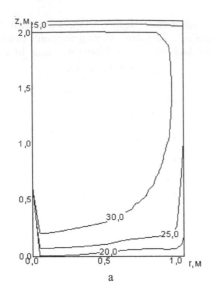
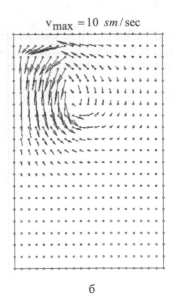

a   б

Figure 4. The temperature field and the flow field two-dimensional situation during stirring at $\tau = 6$ h.

As the result of numerical experiments on mathematical model of anaerobic digestion, biogas yield $V_{b.m}$ was calculated for 1 day with 3 times-a-day stirring. Heater location options are considered across the reactor height only. Heater location height was in the range $z \in (0.01 \div 0.8)H_1$ (Tab. 4).

Table 4. Results of numerical experiments of the influence of natural convection on biogas production rate.

	$z = 0.01H_1$	$z = 0.2H_1$	$z = 0.4H_1$	$z = 0.6H_1$	$z = 0.8H_1$
$V_{bm}, \dfrac{m^3 / m^3}{day}$	1,473	1,46	1,44	1,41	1,4

The similar situation for 1 day with 3 times a day for 20 min stirring was considered to compare. This given situation was calculated using a model based on simplified approach, in which stirring was substituted by equal concentration of all reactants, heat transfer was substituted by a heat balance equation. In this case the mathematical model boils down to a system of differential kinetic equations independent from coordinates. As the result of this model we have (Zemlianka, Gubynskyy 2009):

$$n = 3 \text{ times/day}, \quad V_{bm} = 1{,}53 \; \frac{m^3 / m^3}{day}.$$

Analyzing the results shown in Table 4, we can conclude that in the given operational conditions, a potential biogas production increase for the bioreactor of a small biogas plant by controlling heat and mass transfer processes constitutes up to 5 %. A maximum possible error in biogas yield calculations using simplified approach (Zemlianka, Gubynskyy 2009) as compared with calculations using the mathematical model constitutes up to 10 %.

4 CONCLUSIONS

As a result of mathematical simulations of heat and mass transfer processes in real operational conditions of a small biogas plant, a significant influence of natural convection on the fluid flow and the temperature distribution inside the bioreactor was shown. Considering natural convection and real operational conditions in this given mathematical model, it's possible to get more accurate 5-10% calculations of biogas yeld as compared with calculations by simplified approach. It also allows to choose optimal design parameters for heater installation and optimal characteristics for stirring of biomass.

REFERENCES

Khanal S. 2008. *Anaerobic Biotechnology for bioenergy production. Principles and applications.* A John Wiley & Sons Ltd. Publication: New Delhi: 301.

Zemlianka A.A. 2004. *Mathematical model of hydrodynamics and heat exchange in biogas fermentor with immersed cylindrical heat exchanger* (in Russian). Issues of chemistry and chemical technology, Issue1: 181-187.

Zemlianka O.O. 2007. *Mathematical modeling of the kinetics of organic waste anaerobic fermentation in biogas fermentor* (in Ukrainian). Integrated technologies and energy saving, Issue 4: 47-52.

Zemlianka O.O., Gubynskyy M.V. 2008. *Impact of technological factors on biogas plant performance* (in Ukrainian). Dnipropetrovs'k: Proceedings of the XV International Conference "Heat and Power Engineering in Metallurgy": 101.

Leontyev A.I. 1979. *The theory of heat and mass transfer: a textbook for high schools* (in Russian). Moskow: Vyshaya shcola: 495.

Patancar S. 1980. *Numerical heat transfer and transfer flow.* Taylor & Francis: 197.

Kochubey A.S., Ryadno A.A. 1991. *Numerical modeling of convective transport based on the finite element method* (in Russian). Dnipropetrovs'k: DGU: 227.

Zemlianka O.O., Gubynskyy M.V. 2009. *Choice of rational operation modes biogas plant reactor* (in Ukrainian). Technical thermal physics and industrial power, Issue1: 112-120.

200

Power Engineering, Control and Information Technologies in Geotechnical Systems – Pivnyak, Beshta & Alekseyev (eds)

The electromechanical system of the turning machine tool

A. Bakutin
State Higher Educational Institution "National Mining University", Dnipropetrovs'k, Ukraine

ABSTRACT: The transfer functions of optimal relative to the control signal and feedback control units are obtained, on which the structure and parameters of the correction link in the direct and feedback channel are determined. The cutting speed stabilization system is designed. The simulation of the turning machining process taking into account the elastic links is carried out.

1 INTRODUCTION

Today the engineering industry share of Ukraine is 12%, although in the early 90's it stood at 31% and was the basic industry of the country. Dnipropetrovsk region engineering complex consists of 120 plants, which represent 10% of Ukrainian industry. The main improvement direction of economic development is the creation of the competitive engineering industry, that requires an infusion of new investment and the modernization of machine park. Process automation in engineering and widespread adoption of CNC machines in industry led to the nomination of strict requirements for main motion machine tool electric drives. At the present stage the machine tool industry development is characterized by the transition to continuous speed variation and simplification of the electric drive kinematic structure, that allow to increase productivity and quality of metal through the rational choice of the cutting mode.

2 FORMULATING THE PROBLEM

The requirements for real electromechanical systems are not quite often only to ensure the performance quality by one criterion, such as the speed hold accuracy in the steady state, but by more criteria, such as the control signal response accuracy in dynamic mode, the uniform acceleration value without reference to the load torque, the implementation simplicity and etc., which are determined by the technological features of the control object. Therefore, to build a control system that simulta-

neously satisfies the criterion set, there is a need to optimize it for some objective functions that characterize different aspects of the object that is the multicriteria design of the control system device.

For the predetermined control object characteristics, disturbances, constraints and provided the control object to control device relationship the multicriterion design allows to determine the structure and optimal parameters of the control device. An important condition for the use of this approach is to have algorithms that allow to construct a set of adequate mathematical models for the control devices. The mathematical models of constraints that imposed on the system should provide the selection of the control devices that meet the terms of the performance specification.

3 MATERIALS UNDER ANALYSIS

The speed control system with the feedback has two inputs and one output (Fig. 1). One of the inputs is supplied by the reference signal $g_1(t) + m(t) + n(t)$, which consists of the function that varies according to certain law $g_1(t)$ and the random component $m(t)$ with superimposed on it the noise component $n(t)$. Another input is supplied by the measured control object output signal, which is subtracted from the reference signal. The output of the control device $u(t)$ together with the superimposed on it noise signal, that consists of the regular $g_2(t)$ and the random $f(t)$ components, is fed to the control object input.

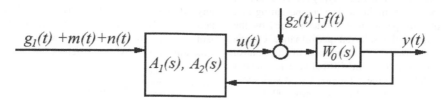

Figure 1.The closed-loop speed control system

The problem solving of the best control device selecting starts with the formation of the possible control device set, from which those are selected that satisfy the imposed performance criteria and constraints. Then the Pareto-optimal set is allocated, from which the final structure and the parameters of the compensating devices are determined for further implementation and analysis. The acceptable control device set boundary is defined, on the one hand, by the control device configuration with maximum difficulty, and on the other hand–by the control device configuration with minimum difficulty. To choose the best control device the analysis of the all possible alternatives within certain boundary, the synthesis of the controllers with predetermined difficulty and the comparison of their performance indexes are carried out.

Assume that the control object is a squirrel cage induction motor. From the classic representation of the induction motor mathematical model with the control by varying the stator supply voltage it can be seen that it contains the cross coupling by the stator current vector components. In case of the compensation or minimization of the cross coupling influence the stator voltage vector component variation can independently set the value of the rotor flux linkage and the motor speed. Then the flux linkage and speed control channels will be divided similar to a DC motor with separate excitation. Assume that the cross coupling by the stator current vector components is compensated and the inner current loop is optimized for the technical criterion, then the control object of the electromechanical system can be described as follows (Bose 2001)

$$W_0(s) = \frac{1.5 \cdot p_{pp} \cdot k_2}{J \cdot k_c \cdot (T \cdot s + 1) \cdot s},$$

where s – Laplace operator; p_{pp} – induction motor pole pairs; $k_2 = {L_m}/{L_2}$ – dimensionless coefficient; L_m – magnetizing inductance, H; L_2 – rotor inductance, H; J – motor inertia moment, $kg \cdot m/s^2$; k_c – stator current sensor coefficient, V/A; T – time constant of the stator loop, s.

The design of the control device with minimum difficulty begins by the device structure configuration choice, and then the transfer function parameters of the control device relative to the back $A_2(s)$ are determined to ensure system stability. Within the selected control device configuration the compensating device construction alternatives are determined by the function $A_2(s)$.Thereafter the control device parameters are evaluated by the selected functional minimization.

To determine the transfer function parameters of the control device relative to feedback $A_2(s)$the modified Hurwitz's criterion is used. The closed-loop performance equation (Zotov 2004):

$$T(s) = P_0(s)G_2(s) + Q_0(s)V_2(s);$$

where

$$A_2(s) = \frac{V_2(s)}{G_2(s)}.$$

For the obtained expression work out a fractional rational function, which includes the polynomial with negative roots $D(s)$,itspower is equal to the power of the performance polynomial $T(s)$,and the coefficients by the highest polynomial terms are equal to each other:

$$\Pi(s) = \frac{T(s)}{D(s)}.$$

The modified Hurwitz's criterion characterizes the vicinity of the polynomial $T(s)$ and $D(s)$coefficients c_i and d_i respectively, according to which the mismatch minimum between them provides the stability of the system (Zotov 2004):

$$I = \sum_{i=0}^{n} \rho_i |d_i - c_i|;$$

whereρ_i – weight coefficients.

The control device design is based on the transfer function configuration and structure relative to the feedback $A_2(s)$. The possible compensating link transfer functions are chosen so, that their structures match the structure of $A_2(s)$. One control device configuration and structure $A_2(s)$ forms a set of possible compensating links. The compensating link transfer function parameters are determined by the minimization of the selected performance criteria similar to the used in the synthesis of the control device with the maximum difficulty. The transfer functions relative to the reference signal and superimposed noise can be written as (Zotov 2004)

$$\widehat{W}(s) = \frac{A_1(s)W_0(s)}{1 + A_2(s)W_0(s)};$$

$$\widetilde{H}(s) = \frac{A_2(s)W_0(s)}{1 + A_2(s)W_0(s)}.$$

Putting restrictions on the system astaticism assumes zero poles of the transfer functions relative to the direct control channel and the feedback channel. To ensure the a_2-th order of the system astaticism relatively to the noise $g_2(t)$ the transfer tion $A_2(s)$ must be of the form

$$A_2(s) = \frac{V_2(s)}{G_2(s)} = \frac{V_2(s)}{s^{a_2}G_2^*(s)},$$

$$W_{1max}(s)$$
$$= \frac{1.389 \cdot 10^{-4} \cdot s^5 + 0.802 \cdot s^4 + 2.311 \cdot 10^3 \cdot s^3 + 3.385 \cdot 10^6 \cdot s^2 + 3.162 \cdot 10^8 \cdot s}{0.032 \cdot s^5 + 180.072 \cdot s^4 + 2.213 \cdot 10^4 \cdot s^3 + 4.996 \cdot 10^5 \cdot s^2 + 5.569 \cdot 10^6 \cdot s - 8.866 \cdot 10^{-9}};$$

$$W_{2max}(s) = \frac{0.49 \cdot s^3 + 732.133 \cdot s^2 + 1.826 \cdot 10^4 \cdot s + 2.276 \cdot 10^5}{1 \cdot 10^{-5} \cdot s^3 + 0.057 \cdot s^2 + 160.651 \cdot s + 2.276 \cdot 10^5};$$

$$W_{1min}(s) = \frac{16{,}0643s + 298.5951}{1.5329s};$$

$$W_{2min}(s) = 0{,}0318.$$

Then the controllers with predetermined difficulty should have the order from the first to the fifth and they exist and will meet necessary requirements. To determine the compensating link transfer functions the control device transfer function relative to the feedback signal should be written

$$A_2(s) = \frac{v_2 s^2 + v_1 s + v_0}{g_2 s^2 + g_1 s},$$

that satisfies the constraints on the first order astaticism relative to the noise $g_2(t)$, since the denominator of the transfer function has a zero root.

The closed-loop control system transfer function is

$$T(s) = v_2 T s^4 + (v_2 + v_1 T)s^3$$
$$+ (kg_2 + v_1 + v_0 T)s^2$$
$$+ (kg_1 + v_0)s + kg_0,$$

then the optimization function can be written as

so the system astaticism is provided by any $V_2(s)$ and $G_2^*(s)$ if the polynomial $V_2(s)$ has no zero roots. To ensure the a_1-th order of the astaticism relative to the reference signal the transfer function $A_1(s)$ must be written as

$$A_1(s) = \frac{V_1(s)}{G_1(s)} = \frac{V_1(s)}{s^{a_1}G_1^*(s)},$$

so by choosing the parameters of polynomials $G_1(s), G_2(s), V_1(s), V_2(s)$ the presence of a_1 zero roots is provided in the polynomials (Zotov 2004)

$$C_1(s)G_2(s)G_1(s)P_0(s)$$
$$+ \big(V_2(s)G_1(s)C_1(s)$$
$$- V_1(s)G_2(s)D_1(s)\big)Q_0(s),$$

and the denominator roots must be negative.

At the earlier stages of research the compensating link transfer functions with the maximum and minimum difficulty relative to the reference and feedback signals were obtained:

$$I = \left|d_3 - \frac{v_2 + v_1 T}{v_2 T}\right| + \left|d_2 - \frac{kg_2 + v_1 + v_0 T}{v_2 T}\right|$$
$$+ \left|d_1 - \frac{kg_1 + v_0}{v_2 T}\right|$$
$$+ \left|d_0 - \frac{kg_0}{v_2 T}\right|,$$

where

$$k = \frac{1.5 \cdot p_{\text{пп}} \cdot k_2}{J \cdot k_{\text{c}}}.$$

The control device evaluation with predetermined difficulty was carried out for the squirrel cage induction motor 4A90L2U3. Taking into account the constraints on the coefficients of the optimization functional

$$d_0 > 0; d_1 > 0; d_2 > 0; d_3 > 0;$$

$$\begin{vmatrix} d_3 & 1 & 0 & 0 \\ d_1 & d_2 & d_3 & 1 \\ 0 & d_0 & d_1 & d_2 \\ 0 & 0 & 0 & d_0 \end{vmatrix} > 0,$$

that introduce constraints to obtain the performance equation roots only from the left side of the complex plane, obtain the control device transfer function relative to the feedback signal:

$$A_2(s) = \frac{316330s^2 + 11.53s - 44.02}{1422s^2 + 1.289s}.$$

After determining the structure and parameters of the function $A_2(s)$, which stabilizes the control object, proceed to the control device design based on its configuration. For the selected block diagram the relations between the control device transfer functions relative to the reference signal $A_1(s)$ and feedback signal $A_2(s)$ and the compensating links $W_1(s)$, $W_2(s)$ are as follows

$$A_1(s) = W_1(s);$$

$$A_2(s) = W_1(s)W_2(s).$$

From the first order astaticism condition relative to the reference signal the control device transfer function $A_1(s)$ should have a zero pole, therefore the possible combinations of the compensating link transfer functions for the determined transfer function $A_2(s)$ can be written as

$$W_1(s) = \frac{a_2s^2 + a_1s + a_0}{b_2s^2 + b_1s};$$

$$W_2(s) = c_0;$$

$$W_1(s) = \frac{a_1s + a_0}{b_1s}; W_2(s) = \frac{c_1s + c_0}{d_1s + d_0}.$$

The next step is to calculate the coefficients of the compensating link transfer functions based on the chosen criteria minimization. Consider the first link configuration alternative. Write the transfer functions of the electromechanical system relative to the reference signal and the noise

$$\widehat{W}(s) = \frac{A_1(s)W_0(s)}{1 + A_2(s)W_0(s)}$$
$$= \frac{k(g_2s^2 + g_1s)(a_2s^2 + a_1s + a_0)}{(Ts + 1)(g_2s^2 + g_1s) + k(v_2s^2 + v_1s + v_0)};$$

$$\widetilde{H}(s) = \frac{A_2(s)W_0(s)}{1 + A_2(s)W_0(s)}$$
$$= \frac{k(v_2s^2 + v_1s + v_0)}{(v_2s^2 + v_1s + v_0)(Ts + 1) + k(g_2s^2 + g_1s)}.$$

The desired transfer function of the regular reference signal component transform $g_1(t)$ is assumed as $U_1(s) = 1$, and the desired transfer function of the regular noise signal component transform $g_2(t)$ is assumed as $U_2(s) = 0$, and for the random com-

ponents – $P_1(s) = 0$ and $P_2(s) = 0$. As the control object contains no positive zeroes and poles, the components that impose the restrictions on the positive zero and pole compensation are not included to the functional. Then the functional for the control device transfer function optimization problem solving relative to the reference signal take the following form

$$I_1 = \frac{1}{2\pi j} \int_{-j\infty}^{j\infty} \left| \left(1 - \widehat{W}(s) \right) \frac{1}{s} \right|^2 ds$$
$$+ \frac{1}{2\pi j} \int_{-j\infty}^{j\infty} \left| s\widehat{W}(s) \right|^2 ds,$$

$$I_2 = \frac{1}{2\pi j} \int_{-j\infty}^{j\infty} \left| \left(1 - \widetilde{H}(s) \right) W_0(s) \right|^2 ds$$
$$+ \frac{1}{2\pi j} \int_{-j\infty}^{j\infty} \left| s\widetilde{H}(s) \right|^2 ds.$$

After the optimization problem is solved the transfer function $A_1(s)$ is obtained

$$A_1(s) = \frac{6793000s^2 + 362.1s - 1382}{1422s^2 + 1.289s}.$$

Based on the expressions of $A_1(s)$ and $A_2(s)$ the compensating link transfer functions for the direct control channel and the feedback channel can be written

$$W_1(s) = \frac{6793000s^2 + 362.1s - 1382}{1422s^2 + 1.289s};$$

$$W_2(s) = 0.0318.$$

For the second compensating link configuration alternative the transfer function synthesis is based on the statement, that the steady-state mode feedback transfer coefficient should ensure the compliance of the output speed value and the input reference signal value:

$$W_2(s) = \frac{c_1s + 10}{d_1s + 314,159},$$

where the control device transfer function relative to the reference signal can be found as follows

$$A_1(s) = \frac{(s - 0.0142)(s + 0.0143)(d_1s + 314.159)}{(s + 0.00091)(c_1s + 10)s}.$$

After the optimization problem is solved the compensating link transfer functions for the direct control channel and the feedback channel are

$$W_1(s) = \frac{3314s^3 + 67930s^2 + 2.947s - 13{,}82}{1.93s^3 + 14.24s^2 + 0.01289s};$$

$$W_2(s) = \frac{1{,}357s + 10}{15{,}319s + 314{,}159}.$$

The cutting machine tool control affects its systems and mechanisms to ensure the certain cutting process parameters for the accuracy, efficiency etc. The structure of the machine tool determines its composition and the relationship between executive devices and auxiliary mechanisms. The control algorithm determines the sequence of the machine control function execution. Depending on the type of the electromechanical transducer and technological requirements for the control performance the different control system organizations can be implemented. In the metal-working manufacturing the main control object is the cutting process, during which all the requirements for the surface quality with maximum performance and minimum cost. The solution to this problem is solved through the use of automatic control systems that provide high precision, high response speed and ensure the quick readjustment of the cutting machine tool.

The purpose of the cutting process control can be the productivity and accuracy increase, the metal cutting cost decrease, the surface smoothness improvement, the maximum technological capability use of the machine tool, cutting tools, electric drives etc. Required parameters of the cutting process quality are achieved by means of the main and feed drive speed variations and the cutting tool position variation position relatively to the workpiece.

The output coordinate of the controlled main motion electric drive is the rotational speed that can be controlled by two ways – stepped speed variation and stepless speed control. By the stepped speed variation one or polyspeed induction motors and mechanical transmissions are applied. This method is obsolete because the modern production imposes stringent requirements on the accuracy and smoothness to the machine tool spindle speed control that is for optimum cutting conditions for increasing productivity and quality of the metal working, as well as for the reduction and simplification of the machine tool kinematic structure. For the stepless speed control the DC and AC electric drives are applied. The variable speed drives allow to automate the process of the speed variation and to set the optimal cutting process through the necessary cutting speed.

The turning machining process is the interaction of the tool, which moves along the workpiece, and the detail that rotates. The relationship between processing parameters is determined by the empirical expression

$$v = \frac{C_v}{T^m t^{x_v} s^{y_v}},$$

where v – cutting speed; C – coefficient that specifies the machining conditions, the workpiece and tool material; T – tool life, t – cutting depth; s – feed; m, x_v, y_v – powers that are depended on the workpiece and cutting tool material properties and metal-working type.

During the turning machining to the tool cutting edge the resisting force is applied, which is decomposed into three mutually perpendicular components: the cutting force that is tangent to the workpiece surface and directed along the direction of the main motion of the spindle; axial force that acts parallel to the axis of the workpiece and is directed opposite to the feed direction; radial force that is directed perpendicular to the axis of the workpiece. Cutting force is calculated by the following empirical expression (Davim 2011)

$$F_z = C_F t^{x_F} s^{y_F} v^n,$$

where C_F – coefficient that specifies the machining conditions, the workpiece and tool material; x_F, y_F, n – powers that are depended on the workpiece and cutting tool material properties and metal-working type.

To describe the functional relationship between the parameters of the cutting process the empirical expressions are used due to the complexity of the physical processes that occur during the cutting of the processed material and the cutting tool.

The main purpose of the cutting mode stabilization systems is to maintain one or more technological parameters according to the selected process control law. In general, automatic control system should provide two processing mode parameter adjustment and include main motion speed andfeed control channels taking into account the external disturbances, the main of which is the cutting depth and hardness of the processed material variation (Fig. 2). The control system consists of the controller C, the main motion drive MMD and the feed drive FD, which have a direct influence on the cutting process CP. Feedback on technological parameters is implemented through the use of sensors S1 and S2, the signal from which is subtracted from the reference main motion speed and feed signals that coming from blocks MMSR and FR respectively. This scheme involves the use of the regulation as the main drive and the feed drive, but often only the one channel stabilization systems are used.

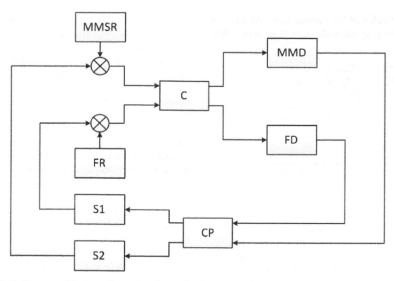

Figure 2. The block diagram of the two channel cutting control system

Consider some typical cutting stabilization systems. Cutting speed stabilization system is one of the most common system and can reduce the cutting time for the workpiece processing, to increase productivity and improve the quality of the workpiece. For example, for the workpiece processing with the variable radius of the surface profile that varies from R_1 to R_2 (Fig. 3) with the constant spindle speed the cutting time can be evaluated as (Granovsky 1985)

$$T_1 = \frac{2\pi L R_1}{9.55 vs},$$

and for the cutting speed stabilization system

$$T_2 = \int_0^L \frac{2\pi r}{9.55 vs} dl,$$

wherecurrent processed surface radius value is evaluated as

$$r = R_2 + l \sin\alpha.$$

Then the cutting time for the workpiece processing can be expressed as

$$T_2 = \int_0^L \frac{2\pi(R_2 + l\sin\alpha)}{9.55 sv} dl$$

$$= \frac{\pi L}{9.55 sv}(R_2 + R_2 + L\sin\alpha)$$

$$= \frac{\pi L}{9.55 sv}(R_1 + R_2).$$

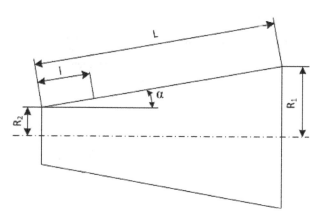

Figure 3. The processed workpiece schematic

The relative value of the cutting time reducing by the use of cutting speed stabilization system calculates as follows

$$\frac{T_1 - T_2}{T_1} = \frac{\frac{2\pi L R_1}{9.55vs} - \frac{\pi L}{9.55sv}(R_1 + R_2)}{\frac{2\pi L R_1}{9.55vs}} = \frac{2R_1 - R_1 - R_2}{2R_1} =$$

$$\frac{1}{2}\left(1 - \frac{R_2}{R_1}\right) = \frac{1}{2}\left(1 - \frac{\omega_{min}}{\omega_{max}}\right) = \frac{1}{2}\frac{D-1}{D},$$

where $\omega_{min} \equiv 1/R_1$; $\omega_{max} \equiv 1/R_2$; $D = \omega_{max}/\omega_{min}$.

The relative reduction of the cutting time to process one workpiece does not depend on the detail length and with the increase of the speed adjustment range is close to 0.5. Besides to the stabilization of the cutting speed the cutting power stabilization systems are widely used, which makes it possible to increase the productivity of the machine tool, the use of the main drive and the cutting tool. The cutting power is determined by measuring the active power consumed by the main drive motor of the machine tool based on the loss rates and gears of the machine tool:

$$P_c = P_m - \Delta P_m - \Delta P_g,$$

where P –cutting power; P_m – active power of the motor; ΔP_m – motor losses; ΔP_g – losses in gears.

Such system is the most effective for the milling machines, which are characterized by a significant change in the length of the cutting tool contact with the workpiece that is determined by the depth and width of the cut. The cutting temperature stabilization system, aimed at the improving the processing performance and stability of the cutting tool, based on the use of the expression for the tool life depending on the temperature in the cutting zone (Davim 2011):

$$T = \left(\frac{C}{\theta}\right)^{\alpha},$$

where C – coefficient that specifies the machining conditions, the workpiece and tool material; θ – cutting temperature; α – powers that are depended on the workpiece and cutting tool material properties and metal-working type.

If the angular speed of the main motion drive remains unchanged, then the turning cutting speed will vary from the maximum at the periphery to the minimum in the center, because of the cutting temperature change will be significant. The cutting temperature stabilization system with decreasing radius of the workpiece automatically increases the angular speed of the main motion drive to preserve constant the set cutting temperature.

In spite of the considered stabilization systems are based on the control of various process parameters, they have generalities and may be equivalent to the results of their work. So for the turning machining with the constant feed the cutting power is defined as

$$P_c = \frac{F_z v}{60},$$

where P_c – cutting power.

So to stabilize the value of the cutting power is necessary to ensure the constant value of the cutting speed, which is similar to the use of the cutting speed stabilizing. At the constant cutting temperature the cutting speed depends on the feed as (Davim 2011)

$$v = \frac{C}{s^k},$$

where k – powers that are depended on the workpiece and cutting tool material properties and metalworking type.

So for the turning manufacturing at the constant values of the feed the maintaining of the constant cutting speed is equivalent to the use of the cutting temperature stabilization system. Block diagram of the cutting speed stabilization system (Fig. 4) contains a cutting speed sensor CSS, which is based on the measurement of the current radius of the workpiece and the angular speed of the main motion motor.

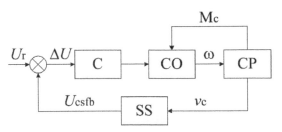

Figure 4. The Block diagram of the cutting speed stabilization system

The dynamic cutting machine tool mathematical model consists of the elastic system, which includes the machine tool, the cutting tool, the workpiece and the working processes such as the cutting, friction and electromagnetic processes in the drivingmotor. The dynamic model is closed and multiloop, the interaction of the system elements is directed and occurs through elastic system that is specified by the separation of theworkflow zones by theelastic system elements. This allows to represent the dynamical system in the form of singleloop machine tool equivalent system that reflects the processes in the driving motor and provides the mechanical system consisting of the elastic system, the processes of the cutting and the friction (Fig. 5).

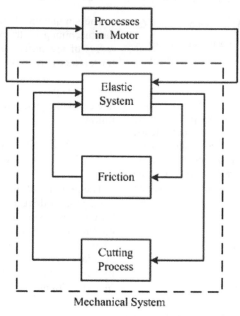

Figure 5. The block diagram of the singleloop equivalent dynamic machine tool

The electromechanical system the main motion machine tool drive represents multimass system consisting of the electric motor rotor, the pulleys, the gears and the spindle holder with machined workpiece. In the study of the drive dynamics the concentrated masses interconnected with instantaneous shafts that reflect the elastic connections between them are considered (Fig. 6).

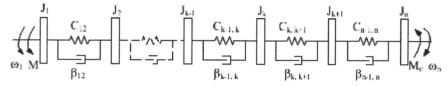

Figure 6. The Diagram of the multimass electromechanical system

The operator equation system for the multimass electric drive system is expressed as follows (Bose 2001)

$$\begin{cases} M - \dfrac{C_{12}}{p}(\omega_1 - \omega_2) - \beta_{12}(\omega_1 - \omega_2) = J_1 p\omega_1 + \beta_1 \omega_1 \\[4mm] \cdots\cdots \\[2mm] \dfrac{C_{k-1,\,k}}{p}(\omega_{k-1} - \omega_k) + \beta_{k-1,\,k}(\omega_{k-1} - \omega_k) - \dfrac{C_{k,\,k+1}}{p}(\omega_k - \omega_{k+1}) - \beta_{k,\,k+1}(\omega_k - \omega_{k+1}) = J_k p\omega_k + \beta_k \omega_k \\[4mm] \dfrac{C_{n-1,\,n}}{p}(\omega_{n-1} - \omega_n) + \beta_{n-1,\,n}(\omega_{n-1} - \omega_n) - M_c = J_n p\omega_n + \beta_n \omega_n \end{cases}$$

where M – motor torque; M_c – load torque; C – elastic coefficient; β – viscosity factor; ω – radial speed; J – inertia of the concentrated rotary mass.

For ease of study, the number of multimass system units is reduced by evaluating of the equivalent elastic system having identical energy performance, through the matching of the elastic coefficients, viscosity factors and inertia of concentrated rotary masses (Bose 2001). After performing the corresponding calculations the equivalent two- or three mass elastic system of the main motion machine tool drive can be obtained (Fig. 7).

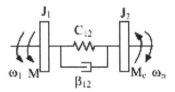

Figure 7. The diagram of the equivalent two mass elastic system

The simulation of the transient processes during the machining of the workpiece with the variable radius of the surface profile (Fig. 3) is carried out for the inertia ratio of the motorrotor and the rotary masses equal to 1/6 and elastic coefficient $C = 400$ kg/s². The transfer function of the control device with predetermined difficult is taken as follow

$$W(s) = \frac{83(0.13s + 1)(0.07s + 1)}{s(0.058s + 1)}.$$

The graphics of the transient processes during the main motion drive launch of the machine tool and turning machining performance of the workpiece confirm the correctness of the constructed cutting speed stabilization system (Fig. 8). It is seen that the angular speed of the motor varies according to a hyperbolic law, and the load torque M_p increases linearly with the workpiece radius increasing.

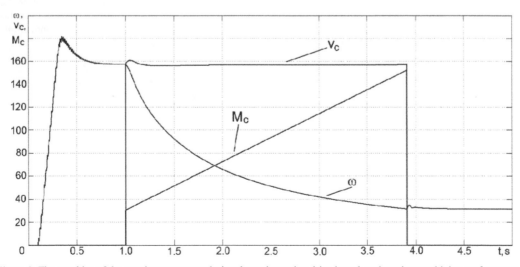

Figure 8. The graphics of the transient processes during the main motion drive launch and turning machining performance

209

4 CONCLUSIONS

The structures and optimal speed control device parameters for the induction motor within the acceptable set of optimal devices, which with the devices with maximum and minimum difficulty, are the initial data for the further multicriteria control system construction are obtained. Based on the technological requirements that apply to the machine tool automatic control and comparative analysis of the stabilization of various technological parameters of the metal processing the cutting speed stabilization system is designed. Using mathematical simulation the correctness of the constructed control device and cutting speed stabilization system to design task is confirmed.

REFERENCES

Zotov M.G. 2004. *Multicriteria design of the automatic control system* (in Russian). Moscow: Binom: Laboratoriya znaniy.

Bose B.K. 2001. *Modern power electronics and AC drives.* Prentice Hall PTR: Upper Saddle River.

Davim J.P. 2011. *Modern Machining Technology.* Cambridge: Woodhead Publishing.

Granovsky G.I. 1985. *Metal Cutting* (in Russian). Moscow: Vyshaya shcola.

Printed and bound by CPI Group (UK) Ltd, Croydon, CR0 4YY

23/10/2024

01778250-0001